REFLECTARRAY ANTENNAS

IEEE Press
445 Hoes Lane
Piscataway, NJ 08854

IEEE Antennas and Propagation Society, *Sponsor*
IEEE APS Liaison to IEEE Press, Robert Mailloux

Technical Reviewers
Richard M. Dickinson, OFF EARTH-WPT
Jeffrey S. Herd, MIT
Yahya Rahmat-Samii, UCLA

REFLECTARRAY ANTENNAS

John Huang

José A. Encinar

IEEE Antennas and Propagation Society, *Sponsor*

IEEE PRESS

WILEY-INTERSCIENCE
A John Wiley & Sons, Inc., Publication

Published by John Wiley & Sons, Inc., Hoboken, New Jersey.
Published simultaneously in Canada.

Wiley Bicentennial Logo: Richard J. Pacifico

Library of Congress Cataloging-in-Publication Data is available.

ISBN: 978-0-470-08491-5

Printed in the United States of America.

10 9 8 7 6 5 4 3 2 1

CONTENTS

Preface ix

Acknowledgments xiii

1. Introduction to Reflectarray Antenna 1

1.1 Description of Reflectarray 1

1.2 Printed Reflectarray 3

1.2.1 Advantages of Reflectarray 3

1.2.2 Disadvantage of Reflectarray 4

References 6

2. Development History 9

2.1 Early Innovations and Developments 9

2.1.1 Waveguide Reflectarray in the 1960s 9

2.1.2 Spiralphase Reflectarray in the 1970s 9

2.1.3 Microstrip Reflectarray in the 1980s 11

2.2 Recent Developments 13

2.3 Comparison with Similar Technologies 20

2.3.1 Array Lens 20

2.3.2 Fresnel-Zone Plate Reflector 22

References 24

3. Antenna Analysis Techniques 27

3.1 Introduction 27

3.2 Overview of Analysis Techniques 29

3.3 Phase-Shift Distribution 34

3.4 Analysis of Rectangular Patches with Attached Stubs 35

3.5 Full-Wave Analysis of Multilayer Periodic Structures 38

3.5.1 Characterization of a Periodic Interface as a Building Block 40

3.5.2 Analysis of Reflectarray Elements in a Periodic Environment 44

3.6 Phase-Shifter Element Based on Single and Stacked Variable-Sized Patches 48

3.6.1 Single-Layer Elements 49

3.6.2 Multilayer Elements 50

3.6.3 Measurements of Phase-Shift and Losses in Waveguide Simulator 52

3.7 Phase-Shifter Element Based on Aperture-Coupled Patches 55

3.7.1 Design of Reflectarray Element 56

3.7.2 Phase Delay Curves 57

3.8 Feed Model and Radiation Patterns 64

3.8.1 Field on the Reflectarray Elements 66

3.8.2 Radiation Patterns 68

3.8.3 Experimental Results 72

3.8.4 Gain Computation 73

References 74

4. Practical Design Approach 79

4.1 Element Effects and Selection 79

4.1.1 Element Reflection Phase 79

4.1.2 Element Beamwidth 81

4.1.3 Element Bandwidth 82

4.1.4 Element Reflection Efficiency 82

4.1.5 Element Spacing 83

4.2 Path Length and Phase Delay Calculation 84

4.3 Radiation Pattern Calculation 85

4.4 Reflectarray Geometry Design 86

4.5 Reflectarray Power Handling 91

References 91

5. Broadband Techniques 93

5.1 Bandwidth Limitation by the Reflectarray Element 93

5.2 Broadband Phase-Shifter Elements 95

5.2.1 Aperture-Coupled Patches 96

5.2.2 Variable-Sized Stacked Patches 97

5.2.3 Other Reflectarray Elements for Bandwidth Improvement 99

5.3 Bandwidth Limitation by Differential Spatial Phase Delay 100

5.4 Broadband Techniques for Large Reflectarrays 104

5.4.1 True Time Delay Reflectarrays 106

5.4.2 Compensation of Phase Delay in a Frequency Band 108

5.4.3 Multifacet Reflectarrays 112

References 116

6. Dual-Band Reflectarray 119

6.1 Dual-Band with a Single-Layer Substrate 119

6.1.1 Circular or Dual-Linear Polarization with Two Widely Separated Frequencies 119

6.1.2 Circular Polarization with Two Closely Separated Frequencies 120

6.1.3 Linear Polarization with Two Widely Separated Frequencies 121

6.1.4 Linear Polarization with Two Closely Separated Frequencies 123

6.2 Dual-Band with Two-Layer Substrates 123

6.2.1 High-Frequency Elements above Low-Frequency Elements 124

6.2.2 Low-Frequency Elements above High-Frequency Elements 125

6.3 Multiband Reflectarray with More than Two Frequencies 131

References 134

7. Recent and Future Applications 137

7.1 Inflatable/Thin-Membrane Reflectarrays 137

7.1.1 X-Band 1-m Inflatable Reflectarray 138

7.1.2 Ka-Band 3-m Inflatable Reflectarray 139

7.2 Contoured Beam Reflectarrays for Space Applications 143

7.2.1 Pattern Synthesis 145

7.2.2 Practical Designs 150

7.3 Multi-Beam Reflectarrays 169

7.4 Amplifying Reflectarray 177

7.5 Folded Compact Reflectarray 182

7.6 Cassegrain Offset-Fed Configurations 190

7.7 Very Large Aperture Applications 194

7.8 Beam Scanning Reflectarrays 195

References 201

Index **207**

PREFACE

The concept of the reflectarray antenna was introduced in 1963 using waveguide elements, but the real interest in reflectarrays only came about in the late 1980s with the development of low-profile printed antennas. For this reason, the printed reflectarray can be considered as a fairly new type of antenna. A reflectarray is made up of an array of radiating elements that provide a preadjusted phasing to form a focused beam when it is illuminated by a feed, in a similar way to a parabolic antenna. Printed reflectarrays combine certain advantages of reflector antennas and phased arrays. They are manufactured on a planar substrate using printed circuit technology and offer the possibility of beam steering as phased arrays; on the other hand, the feeding mechanism (as in a reflector antenna) eliminates the complexity and losses of the feeding network used in planar arrays, thus providing a higher efficiency. Reflectarrays have demonstrated their capability to produce contoured beams, which are conventionally generated by using shaped reflectors or phased arrays. Recently, some potential applications of reflectarrays in space have been researched, such as contoured beam antennas for Direct Broadcast Satellites and very large inflatable antennas. However, there is one major shortcoming of the reflectarray, which is its narrow-bandwidth behavior, but the bandwidth has been significantly increased in recent developments.

The purpose of this book is to present a comprehensive overview of reflectarray antennas, including the operating principles, their advantages over other antennas, their development history, analysis techniques, practical design procedures, bandwidth issues, and wideband techniques, as well as their applications and recent developments. This book can be used as a reference book for graduate students, researchers, and antenna engineers. Furthermore, it will allow the reader to become more familiar with this relatively new type of antenna and will provide valuable support in designing these antennas.

The book is organized into seven chapters. Chapter 1 presents a general introduction to reflectarray antennas, including their operating principles, the most common implementations, their most significant features, and a description of the advantages and drawbacks with respect to other types of antenna. A detailed development history of the reflectarray since its invention is presented in Chapter 2. It will greatly enhance the ability of an engineer to understand the reflectarray system if he is familiar with the evolution of the reflectarray antenna. In the same chapter, performance comparisons with two similar technologies, array lens and Fresnel-Zone plate reflector, are also briefly discussed.

The reflectarray is a relatively complex antenna, and an accurate analysis technique is essential for precise predictions of the radiation features, such as efficiency, gain, co- and cross-polar radiation patterns, and bandwidth. Chapter 3 provides a detailed discussion of the different approaches used for the analysis of reflectarrays. A full-wave technique based on the Method of Moments in the Spectral Domain and Floquet modal expansions, under the assumption of local periodicity to account for mutual couplings, is described in detail. Although the method of analysis has already been described in journal papers, in this chapter the method is focused to the specific analysis of reflectarray antennas. The analysis technique is used to compute the phase response and losses of different types of reflectarray elements, such as printed patches with attached or aperture-coupled stubs and varying-sized patches in single- and multiple-layer configurations. Several results are presented to validate the analysis tool and to show the capabilities and limitations of each element type. Finally, the chapter describes a technique for the computation of co- and cross-polar radiation patterns, including the modeling of the feed-horn.

Chapter 4 is devoted to providing practical skills for the design of the reflectarray antenna. Apart from the analysis technique, there are some aspects that must be taken into account during the design of a reflectarray antenna, such as the selection criteria for the phasing elements, the appropriate spacing to provide enough phase-range and avoid the appearance of grating lobes, the geometrical definition of the antenna, etc. These aspects are essential to achieving a good efficiency and are discussed in Chapter 4.

The most severe drawback in reflectarray operation is its narrow frequency band, and much effort has been made in recent years to overcoming this limitation. Chapter 4 is devoted to bandwidth improvement in reflectarrays. The bandwidth in reflectarrrays is mainly limited by two different factors: the phase response of the radiating element and the different path lengths from the feed to the phase front. After discussing the two factors in detail, several solutions are presented in Chapter 5 to improve the bandwidth. Two different types of broadband reflectarray elements are analyzed: one based on stacked patches of varying size and the other using patches with aperture-coupled lines. A 16 percent bandwidth was achieved by using two stacked patches as reflectarray element. On the other hand, the effect of different path lengths is only significant in large reflectarrays as in the case of antennas for space applications. Several techniques are described to overcome the bandwidth limitation produced by the different path lengths and some results are given for large antennas. The first technique is based on the implementation of delay lines aperture-coupled to printed patches to compensate for the real phase delay in the whole range (several times 360°), the second one consists of compensating the spatial phase delay in a given frequency band with the phase of the reflection coefficient, and the last one, more suitable for very large apertures, uses a faceted configuration that approximates the shape of a parabolic surface.

Chapter 6 presents dual-band and multi-band techniques for a single reflectarray to handle multiple frequencies that are separated far apart.

Several important recent applications, as well as possible future applications, are presented in the final chapter. Examples such as inflatable reflectarray, contour-beam applications, multibeam reflectarrays, amplifying reflectarrays, a folded low-profile configuration, a Cassegrain offset configuration, very large aperture applications, and beam scanning reflectarrays are presented in some detail. Due to the multitude of capabilities, the development and application of reflectarrays are expected to carry on during the next decades.

ACKNOWLEDGMENTS

The work presented in this book was supported in part by the following institutions:

National Aeronautics and Space Administration (NASA)
European Space Agency (ESA)
Jet Propulsion Laboratory (JPL), California Institute of Technology
Spanish Commission of Science and Technology (CICYT)
Mexican National Council of Science and Technology (CONACYT).

The authors wish to express their gratitude to E. Carrasco and M. Arrebola, from Universidad Politecnica de Madrid, to Dr. Agustin Zornoza, from EADS Astrium Ltd, to Dr. Herve Legay from Thales Alenia Space, and to Dr. Leri Datashvili from Technical University of Munich, for providing some material presented in this book. Special thanks are expressed to Professor W. Menzel of University of Ulm (Germany) for making available some pictures and results of his folded reflector antennas. Special thanks are also extended to Dr. R. J. Mailloux of USAF for his encouragement and suggestion of writing this book.

CHAPTER 1

Introduction to Reflectarray Antenna

For most radar and long distance communications, the need for high-gain antennas is unavoidable. Traditionally, high-gain applications have relied upon parabolic reflectors or arrays [1]. However, the parabolic reflector in many cases, due to its specifically curved surface, is difficult to manufacture, in particularly at higher microwave frequencies. It also lacks the ability to achieve wide-angle electronic beam scanning. On the other hand, the high-gain array antenna, when equipped with controllable phase shifters, can achieve wide-angle beam scanning electronically, but generally becomes very expensive due to its complicated beamformer and many high-cost amplifier modules. The amplifier modules must be used to alleviate the problem associated with the power inefficiency that occurs in the high-loss beamformer and phase shifters. As a result, a third type of antenna, namely the "reflectarray", has evolved to mitigate the disadvantages associated with either the parabolic reflector or the conventional array.

1.1 DESCRIPTION OF REFLECTARRAY

The reflectarray [2, 3] is an antenna consisting of either a flat or a slightly curved reflecting surface and an illuminating feed antenna as shown in Fig. 1.1. On the reflecting surface, there are many radiating elements (e.g., open-ended waveguides, printed microstrip patches, dipoles, or rings) without any power division transmission lines. The feed antenna spatially illuminates these reflectarray elements that are predesigned to reradiate and scatter the incident field with electrical phases that are required to form a planar phase front in the far-field distance. In other words, the predesigned phases of all elements are used to compensate for the different phases associated with the different path lengths ($S_1, S_2, \ldots, S_n$ in Fig. 1.1) from the illuminating feed. This operation is similar in concept to the use of a parabolic reflector that utilizes its unique curvature to reflect and form a planar phase front when a feed is placed at its focal point. Thus, the term "flat reflector" is sometimes used to describe the

Reflectarray Antennas, by John Huang and José A. Encinar

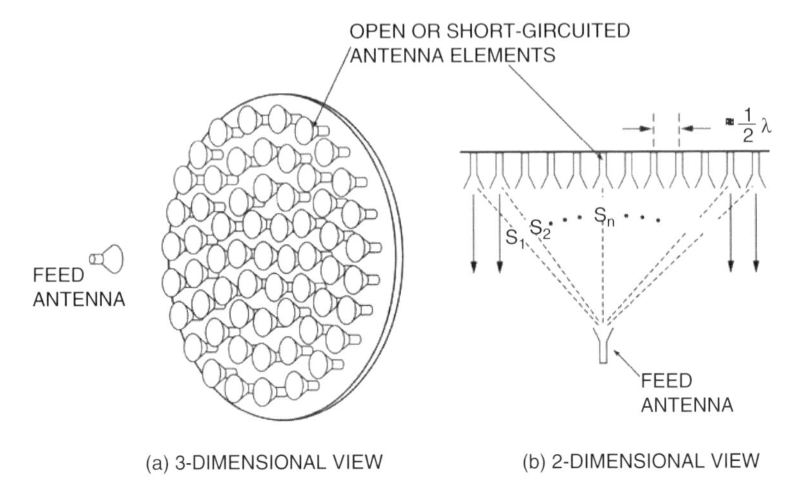

Figure 1.1. Configuration of a reflectarray antenna.

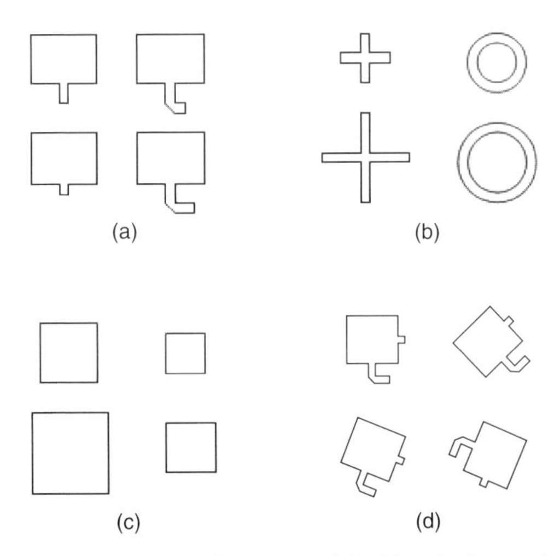

Figure 1.2. Various reflectarray elements, (a) identical patches with variable-length phase delay lines, (b) variable-size dipoles or loops, (c) variable-size patches, (d) variable angular rotations.

reflectarray, which utilizes both technologies of reflector and array. As shown in Fig. 1.2, there are several methods for reflectarray elements to achieve a planar phase front. One is to use identical microstrip patches with variable-length phase delay lines attached [4, 5] so that they can compensate for the phase delays over the different paths from the illuminating feed. Another is to use variable-size patches, dipoles, or rings [6–8] so that elements can have different scattering impedances and, thus, different phases to compensate for the different feed-path delays. With the third method, for circular polarization only, the reflectarray has all identical circularly polarized elements but with different angular rotations [9] to compensate for the feed path-length differences.

1.2 PRINTED REFLECTARRAY

Reflectarrays using printed microstrip elements have been developed to achieve low reflecting surface profile, small antenna mass, and low manufacturing cost. A configuration of a reflectarray using printed patch elements with variable-length delay lines is shown in Fig. 1.3. These printed reflectarrays combine some of the salient features of the traditional parabolic reflector antenna and the microstrip array technology. Its advantages, as well as disadvantages, when used as a large-aperture antenna are separately discussed below.

1.2.1 Advantages of Reflectarray

Similar to a parabolic reflector, the reflectarray can achieve very good efficiency (>50 percent) for a very large aperture since no power divider is needed and thus very little resistive insertion loss is encountered here. On the other hand, very similar to an array antenna, the reflectarray can have its main beam designed to tilt at a large angle (>50°) from its broadside direction. Low-loss electronic phase shifters can be incorporated into the elements for wide-angle

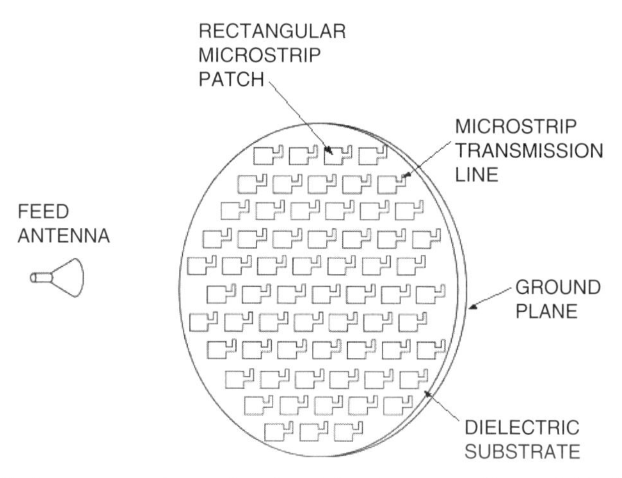

Figure 1.3. Reflectarray using printed patch elements with variable-length delay lines.

electronic beam scanning [10, 11]. With this beam scanning capability of the reflectarray, the complicated high-loss beamforming network and high-cost transmit/receive (T/R) amplifier modules of a conventional phased array are no longer needed.

One significant advantage of the printed reflectarray is that, when a large aperture (e.g., 10-m size) spacecraft antenna requires a deployment mechanism, the flat structure of the reflectarray allows a much simpler and reliable folding mechanism compared with that required for the doubly curved surface of a parabolic reflector. In addition, an inflation system augmented with a large, flat, thin membrane reflectarray aperture can be deployed using a rolling mechanism to form an inflatable antenna. The flat reflecting surface of the reflectarray also lends itself for flush mounting onto an existing flat structure without adding significant amount of mass and volume to the overall system structure. A reflectarray with hundreds or thousands of elements, being in the form of a printed microstrip antenna, can be fabricated with a simple and low-cost chemical etching process, especially when produced in large quantities.

Another major feature of this antenna is that, with a large number of elements in a reflectarray having elemental phase adjustment capability, it can achieve very accurate contour beam shape by using a phase synthesis technique [12, 13]. Similar to the parabolic reflector, multiple-beam capability can also be achieved by placing multiple feed elements at the focal area of the antenna. The reflectarray technology can be applied throughout the microwave spectrum, as well as at the millimeter-wave frequencies.

1.2.2 Disadvantage of Reflectarray

With all the above capabilities, there is one distinct disadvantage associated with the reflectarray antenna. This is its inherent characteristic of narrow bandwidth, which generally cannot exceed much beyond ten percent depending on its element design, aperture size, focal length, etc. The bandwidth performance of a reflectarray [14, 15] is no match to that of a parabolic reflector, where theoretically infinite bandwidth exists. For a printed microstrip reflectarray, its bandwidth is primarily limited by two factors. One is the narrow bandwidth of the microstrip patch elements on the reflectarray surface and the other is the differential spatial phase delay.

1.2.2.1 Bandwidth Limited by Element. The microstrip patch element generally has a bandwidth of about 3 to 5 percent. To achieve wider bandwidth for a conventional microstrip array, techniques such as using thick substrate for the patch, stacking multiple patches [16, 17], and using sequentially rotated subarray elements have been employed. More than 15 percent bandwidths have been reported.

1.2.2.2 Bandwidth Limited by Differential Spatial Phase Delay. The second reflectarray limiting factor, the differential spatial phase delay, can be

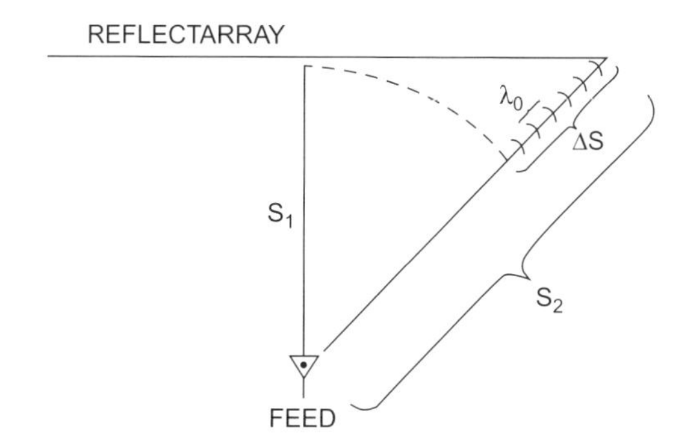

Figure 1.4. Differential spatial phase delay of reflectarray.

best explained by referring to Fig. 1.4 where the differential spatial phase delay, ΔS, is the phase difference between the two paths S1 and S2 from the feed to the reflectarray elements. This ΔS can be many multiples of the wavelength (λ) at the center operating frequency. It can be expressed as $\Delta S = (N + d)\lambda$ where N is an integer and d is a fractional number of a free-space wavelength λ. At each element location, d is compensated by an appropriate phase delay achieved by the reflectarray element design (achieved by variable patch size, variable phase delay line length, etc.). As the frequency changes, the factor $(N + d)\lambda$ becomes $(N + d)(\lambda + \Delta\lambda)$. Since the design and the compensating phase for each element are fixed for the center frequency, a frequency excursion error will occur in the reradiated phase front. The amount of phase change in each path when compared with a reference path, say S1, is $(N + d)\Delta\lambda$ which can be a significant portion of a wavelength or 360°.

To reduce the amount of frequency excursion error, the integer number N must be reduced. There are several methods to reduce N. One is to design the reflectarray with a larger focal-length-to-diameter (f/D) ratio and hence to minimize the difference between paths S1 and S2. The second way is simply to avoid the use of a reflectarray with a large electrical diameter. The third method to reduce frequency excursion error is to use time delay lines or partial time delay lines instead of the phase delays. In other words, when using the phase delay line technique (not the variable patch size technique), instead of using $d\Delta\lambda$ for the delay line length, $(N + d)\Delta\lambda$ could be used for the delay line. Certainly, additional line insertion loss and needed real estate for the lines are issues to be encountered.

Another method to increase the bandwidth is to use, instead of a complete flat reflectarray surface, a concavely curved reflectarray with piecewise flat surfaces. This piecewise flat reflectarray has advantages over a curved parabolic reflector: its beam is able to be scanned to large angles with phase shifter inserted into each element, and, for a space deployable antenna, the piecewise flat surfaces in some cases are more easily folded into a smaller stowed volume.

The narrow bandwidth behavior of the reflectarray will be discussed further in Chapter 5. Techniques to expand the bandwidth will also be presented. Although the reflectarray has bandwidth limitation, due to its multitude of capabilities, the development, research, and application of the printed reflectarray antenna would remain to be boundless in the future.

This book is divided into seven chapters covering different aspects of the reflectarray. Chapter 2 gives a detailed development history of the reflectarray since its invention. It would greatly enhance the ability of an engineer in understanding the reflectarray system if he or she is familiar with the evolution of the reflectarray antenna. In this same chapter, performance comparisons with two similar technologies, array lens and Fresnel-Zone plate reflector, are also briefly discussed. Reflectarray is a relatively complex antenna. Its radiation efficiency is highly dependent on the accuracy of the analysis technique and design knowledge. Chapter 3 provides a detailed discussion on several viable analysis techniques, while Chapter 4 gives practical design points for the element and antenna geometry designs. Since the major drawback of the reflectarray is its relatively narrow bandwidth, several researchers have developed techniques to broaden its bandwidth from a few percent to more than ten percent. Chapter 5 discusses the bandwidth limitation of a reflectarray and its broadband techniques. Chapter 6 presents dual-band and multi-band techniques for a single reflectarray aperture to handle multiple frequencies that are widely separated. Although the reflectarray was invented more than 40 years ago, its application has not been diversified until recently due to the development of the printable microstrip reflectarrays. Several important recent applications, as well as possible future applications, are presented in the final Chapter 7. Examples such as inflatable reflectarray, contour-beam application, multibeam reflectarray, amplifying reflectarray, folded low-profile configuration, Cassegrain offset configuration, very large aperture application, and beam scanning reflectarray are presented with some details. Due to the multitude of capabilities, the development and application of reflectarrays are expected to continue well into the future.

REFERENCES

1. H. Jasik, *Antenna Engineering Handbook*, Chapters 12 and 15, MaGraw-Hill, New York, 1961.
2. D. G. Berry, R. G. Malech, and W. A. Kennedy, "The reflectarray antenna," *IEEE Trans. Antennas Propagat.*, Vol. AP-11, Nov. 1963, pp. 645–651.
3. H. R. Phelan, "Spiralphase reflectarray for multitarget radar," *Microwave Journal*, Vol. 20, July 1977, pp. 67–73.
4. R. E. Munson and H. Haddad, "Microstrip reflectarray for satellite communication and RCS enhancement and reduction," U.S. patent 4,684,952, Washington, D.C., August 1987.

5. J. Huang, "Microstrip reflectarray," IEEE AP-S/URSI symposium, London, Canada, June 1991, pp. 612–615.
6. D. M. Pozar and T. A. Metzler, "Analysis of a reflectarray antenna using microstrip patches of variable size," *Electronics Letters*, April 1993, pp. 657–658.
7. A. Kelkar, "FLAPS: conformal phased reflecting surfaces," *Proc. IEEE National Radar Conf.*, Los Angeles, California, March 1991, pp. 58–62.
8. Y. T. Gao and S. K. Barton, "Phase correcting zonal reflector incorporating rings," *IEEE Trans. Antennas Propagat.*, Vol. 43, April 1995, pp. 350–355.
9. J. Huang and R. J. Pogorzelski, "A Ka-band microstrip reflectarray with elements having variable rotation angles," *IEEE Trans. Antennas Propagat.*, Vol. 46, May 1998, pp. 650–656.
10. J. M. Colin, "Phased array radars in France: present and future," IEEE symposium on Phased Array System and Technology, Boston, Massachusetts, October 1996, pp. 458–462.
11. A. A. Tolkachev, V. V. Denisenko, A. V. Shishlov, and A. G. Shubov, "High-gain antenna system for millimeter-wave radars with combined electrical and mechanical beam steering," IEEE symposium on Phased Array System and Technology, Boston, Massachusetts, October 1996, pp. 266–271.
12. D. M. Pozar, S. D. Targonski, and R. Pokuls, "A shaped-beam microstrip patch reflectarray," *IEEE Trans. Antennas Propagat.*, Vol. 47, July 1999, pp. 1167–1173.
13. J. A. Encinar and J. A. Zornoza, "Three-layer printed reflectarrays for contour beam space applications," *IEEE Trans. Antennas Propagat.*, Vol. 52, May 2004, pp. 1138–1148.
14. J. Huang, "Bandwidth study of microstrip reflectarray and a novel phased reflectarray concept," IEEE AP-S/URSI symposium, Newport Beach, California, June, 1995, pp. 582–585.
15. D. M. Pozar, "Bandwidth of reflectarrays," *Electronics Letters*, Vol. 39, Oct. 2003, pp. 1490–1491.
16. J. A. Encinar, "Design of two-layer printed reflectarray using patches of variable size," *IEEE Trans. Antennas Propagat.*, Vol. 49, October 2001, pp. 1403–1410.
17. J. A. Encinar and J. A. Zornoza, "Broadband design of three-layer printed reflectarrays," *IEEE Trans. Antennas Propagat.*, Vol. 51, July 2003, pp. 1662–1664.

CHAPTER 2

Development History

2.1 EARLY INNOVATIONS AND DEVELOPMENTS

2.1.1 Waveguide Reflectarray in the 1960s

The reflectarray antenna concept was first conceived during the early 1960s by Berry, Malech, and Kennedy [1]. Short-ended waveguide elements with variable-length waveguides, as illustrated in Fig. 2.1, were used to demonstrate the capability of achieving co-phasal reradiated far-field beams. The electromagnetic waves from the feed horn illuminate and couple into the waveguides from the open ends and travel as in transmission lines down to the other shorted ends. The shorted ends reflect all signals and reradiate out from the open ends. By controlling the lengths of individual waveguide elements, the phases of the reradiated signals could be appropriately adjusted to form a desired beam in the far-field distance. Because most wireless operations during this early time were done at relatively low microwave frequencies, the large-waveguide reflectarrays resulted in very bulky and heavy antennas, and thus this antenna concept was not pursued until more than ten years later. In addition, the efficiencies of these reflectarrays were not studied and optimized.

2.1.2 Spiralphase Reflectarray in the 1970s

In the mid-1970s, a very clever concept of "spiraphase" reflectarray was developed by Phelan [2], where switching diodes, as illustrated in Fig. 2.2, were used in a four-arm spiral or cross-dipole element of a circularly polarized reflectarray to electronically scan its main beam to large angles from the broadside direction. This is possible because, by angularly rotating a circularly polarized radiating element, its propagating electrical phase will also change by an appropriate amount proportional to the amount of rotation. By switching the diodes to activate different pairs of the spiral arms, not only can the phase be adjusted to form a co-phasal beam in the far-field, but it is also possible to electronically scan the beam to wide angles. However, due to the thick spiral

Reflectarray Antennas, by John Huang and José A. Encinar

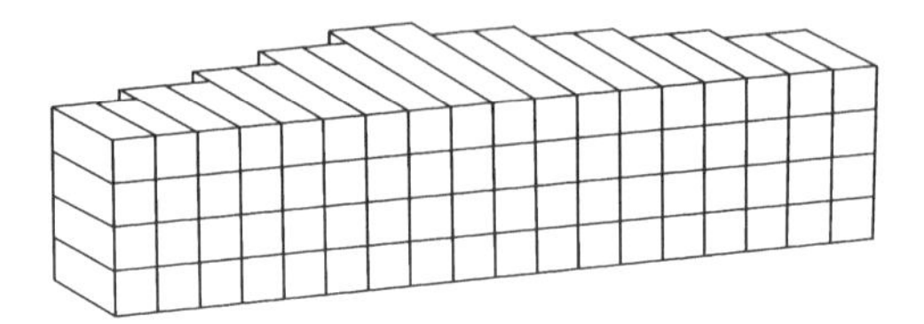

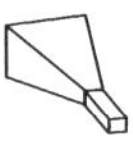

Figure 2.1. Conceptual drawing of the very early reflectarray using open-ended waveguides as elements.

Figure 2.2. Reflectarray using a 4-arm spiral as element with switching diodes at center to achieve a 2-bit phase-shift system for circular polarization.

cavity (quarter-wavelength depth) and large electronic components (diode biasing circuitry), the spiraphase reflectarray was still somewhat bulky and heavy. Its aperture efficiency was still relatively poor in particular due to its large element size and spacing, which are inherent characteristics of the spiral-arm radiator. In particular, at the higher end of the spiral frequency band, the element spacing could be significantly larger than a half free-space wavelength, which could lead to undesirable grating lobes and thus poor efficiency. As a result, no continued development effort was followed in the next ten years. It should be noted here that, in order to have good efficiency for the reflectarray, the intricate relations between the element beamwidth, element spacing, and focal length/diameter (f/D) ratio must be well designed; otherwise, a large backscattered component field or a mismatched surface impedance would result. The relationship between efficiency and reflectarray design parameters will be further discussed in Chapter 4.

2.1.3 Microstrip Reflectarray in the 1980s

Due to the introduction of the printable microstrip antennas, the technologies of combining the reflectarray and microstrip radiators were investigated, and a typical configuration is illustrated in Fig. 2.3. The first mention of using microstrip elements for reflectarray was by Malagisi [3] in 1978. During the same year, the first attempt to analyze the microstrip reflectarray element using infinite array approach was carried out by Montgomery [4]. But there was no further development of the microstrip reflectarray until about a decade later. In the late 1980s and early 1990s, various printed microstrip reflectarray antennas were developed for the purpose of achieving reduced antenna size and mass. These printed reflectarrays came in various forms as shown in Fig. 1.2 and all have flat, low-profile, and low-mass reflecting surfaces. The ones that used identical patch elements with different-length phase delay lines or open-circuited microstrip stubs [5–10] have their elements similar to those shown in Fig. 1.2(a). The phase delay stubs, having lengths in the order of half-wavelength or less, are used to compensate for the phase differences of the different path lengths from the illuminating feed. Assuming the stub line is electrically matched to the patch element, the feed horn signal received by the patch antenna enters the stub line, is reflected by the stub's open circuit (which has a reflection coefficient of 1), and re-enters the patch where it is reradiated

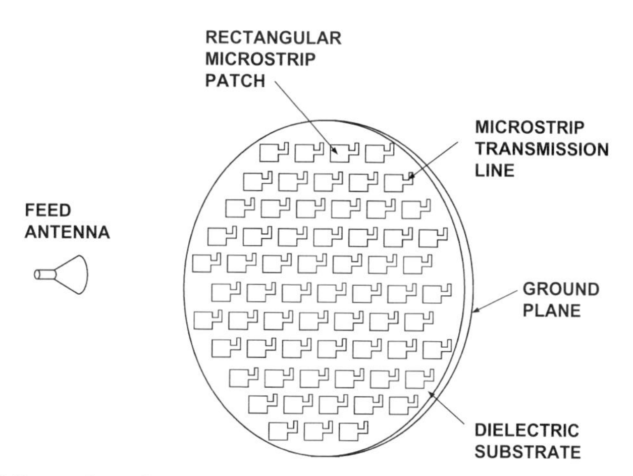

Figure 2.3. Microstrip reflectarray with identical patches but different-length phase delay lines.

with the accumulated phase shift from the stub line. With all elements' stub-line lengths appropriately designed, a co-phasal far-field beam can be formed by the reflectarray.

The second approach, shown in Fig. 1.2(b), uses elements that are made of printed dipoles with variable dipole lengths [11]. Different dipole lengths will yield different scattering impedances, which then provide the different phases needed to compensate for the different path-length delays. Similarly, microstrip patches with variable patch sizes [12], shown in Fig. 1.2(c), were also developed. The reflectarray in this development was shown to provide equivalent efficiency performance to an identical reflectarray based on using the variable-length tuning stubs. However, the variable patch-size reflectarray does exhibit superior cross-polarization performance due to the lack of leakage radiation from the stubs. A technique was also developed for the patches using variable-length tuning stubs to minimize cross-polarization [10]. With this technique, the reflectarray aperture is separated into four quadrants with mirror symmetry arrangement as shown in Fig. 2.4, where square patch elements with reverse orientations and reverse phases for the four quadrants are used to suppress cross-pol radiations.

The development of the variable-patch-size technique also led to the detailed design and analysis of the reflectarray by using the Method of Moments (MoM) technique augmented with the infinite array theory [13, 14]. It is believed that one disadvantage of variable-size approach compared to the variable-stub-length approach is its limited realizable phase range. Usually, the phase range attainable using this technique is somewhat less than 360°. As a result, there is always an unattainable phase range that leads to phase errors in the reflectarray's surface. This variable-size approach, as well as the variable-

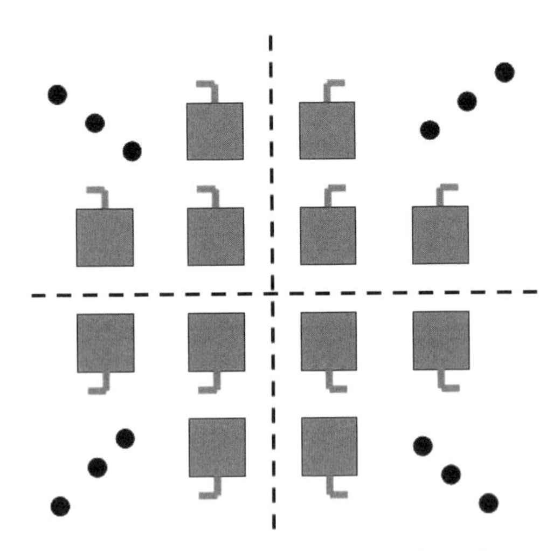

Figure 2.4. Cross-polarization reduction by reverse orientations and reverse phases.

length delay-line approach, may lead to increased specular reflection for off-broadside incident rays because of the large differences in radar cross-sections (RCS) of the constituent elements. To achieve a phase variation close to 360°, the patch size needs to be significantly changed with a total variation of about 40 percent. For the off-resonant patch sizes, a certain amount of specular reflection will occur for off-broadside incident rays, which will lead to inefficiency.

The concept of using circularly polarized microstrip patches with identical size but variable angular rotations [15, 16], shown in Fig. 1.2(d), to form a co-phasal far-field reflectarray beam was conceived and developed. In this development, both the patch element size and attached stub-line length were fixed in dimensions. By physically rotating the element, a phase shift is achieved similar to a circularly polarized (CP) phased array using rotatable elements, the difference being that the phase shift applied to the reflectarray signal is doubled because the reflection process. In other words, for example, an angular rotation of 30° will yield a phase change of 60° for the reflectarray element. This approach achieved superior performance in terms of sidelobe and cross-polarization levels [16] due to an averaging effect of large number of "pseudo-randomly" arranged elements. It is believed that this approach could also achieve better efficiency due to the lack of specular reflection for off-broadside incident rays. This is because all elements are now identical and resonating at the same frequency, and thus have the same RCS regardless of angular rotation even for off-broadside incident rays.

2.2 RECENT DEVELOPMENTS

In addition to those approaches shown in Figs. 2.1 to 2.3, several other reflectarrays or equivalent developments during the 1990s are worth mentioning here. Printed variable-length dipole elements, shown in Fig. 2.5, were used to form a frequency scanned grating-reflector antenna with an offset feed [17]. When the frequency changes, with an offset feed, change in element spacing will lead to a phase change for each row of elements in the offset plane and thus form a different beam position in this offset plane. For the nonoffset plane, the right and left sides of the waves will tilt the same amount and the net result will be a centered beam with broadened beamwidth and lowered gain. Printed annular rings of variable diameters arranged in Fresnel-Zone configuration, as shown in Fig. 2.6, were also used to focus the beam [18]. Within each annular zone of the antenna where the path-length difference varies one full wavelength (360°), all elements are identical in size. In different zones, the ring sizes are different and are appropriately designed to have scattering phases forming a concentrated main beam in the far-field distance. This is not an efficient reflector since the required phases are discretized from the different zones and are not continuous as in the case of the reflectarray. The characteristics of this Fresnel-Zone reflector will be further discussed later in this section.

Figure 2.5. Printed dipole frequency-scanning grating reflector antenna. (From [17], © Sept. 1990 IEEE.)

Figure 2.6. Printed Fresnel-Zone antenna with variable-diameter annular rings.

At the 1996 Phased Array Conference, a 94-GHz monolithic reflectarray [19] fabricated in a single waffle, using one-bit PIN diode phase shifters, was reported to achieve wide-angle (±45°) electronic beam scanning. Dipole antenna elements were printed, using integrated circuit lithography process, on a silicon passivated wafer with built-in PIN diodes. A second wafer, containing the diode driving circuits, is linked at the back of the reflectarray through bumps. At the same conference, a 35-GHz reflectarray, using waveguide/dielectric elements with three-bit ferrite phase shifters [20], was also reported to achieve ±25° beam scanning. The reflectarray has a hexagonal shape with 0.64-m diameter and contains 3600 waveguide-dielectric elements having element spacing about 1.1 free-space wavelength. Each element has a flat-topped pattern with sharp edge cut-off to suppress grating lobes. Electronic beam steering is provided by latched ferrite phase shifters of the Faraday type, which are assembled in subarrays combined with biasing control cells. This reflectarray is capable of operating with right-hand circular polarization (RHCP) on transmission and with left-hand circular polarization (LHCP) on reception. A more recent electronic beam scanning reflectarray development uses voltage-controlled Varactor diodes [21, 22], where only one or two control lines are needed to achieve beam scanning with relatively lower loss. This Varactor-diode approach will be discussed in more detail in Chapter 7. The potential of achieving even lower phase shifter loss by using the micro-electro-mechanical (MEM) switches is currently under development [23].

One proposed technique [15], although not yet fully developed, is worth mentioning here. By using the angular rotation technique with circularly polarized elements, as depicted in Fig. 2.7, miniature or micro-machined motors could be placed under each element to achieve wide-angle beam scanning without the need of expensive T/R modules and phase shifters. With this approach, nearly all the major Radio Frequency (RF) losses are eliminated. This concept was later experimentally demonstrated by placing a two-element

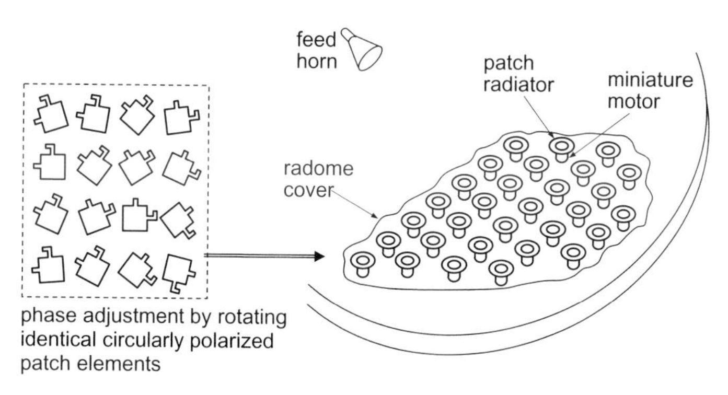

Figure 2.7. Miniature micro-machined motors used to achieve beam scan by employing the angular rotation technique with CP elements.

array with miniature motors in a waveguide simulator [24] and by a small five-element array [25]. A very clever technique of using a miniature motor to rotate a single dipole to achieve CP beam scanning was recently proposed [26]. This dipole must be spaced with an appropriate distance from the ground plane for a CP wave to form.

Another useful application of the reflectarray surfaced in the late 1990s. It used printed-reflectarray elements on thin membranes for spacecraft inflatable antenna application. A deployable and low-mass 1-m diameter inflatable reflectarray antenna [27] at the X-band frequency was successfully developed for technology demonstration. A large inflatable antenna is ideal for application in space where it is deadly calm, and it can significantly reduce the stowage volume in the spacecraft launch vehicle and reduce its mass to minimize fuel consumption. Another unique spacecraft application of the reflectarray was conceived [28] and developed [29] by using its many elements, with a numerical phase synthesis technique, to form a uniquely shaped contour beam. Shaped contour beams can be used in satellite applications to provide a specified coverage area on the surface of the Earth. From all the above developments, it can be seen that, at the end of the 20th century, the reflectarray antenna technology was becoming mature enough and had a variety of possible applications throughout the microwave and millimeter-wave spectra.

In the early 2000s, the development of reflectarray has grown rapidly and several performance improvement and innovative techniques are worth mentioning here. One used multi-layer stacked patches, as shown in Fig. 2.8, to improve the reflectarray bandwidth from a few percent to more than ten percent [30, 31]. A stacked patch structure can yield a smooth phase-versus-size characteristic, and more importantly, can yield phase ranges far in excess of 360°. In particular, by varying the dimensions of three stacked patches, over 600° of phase range was demonstrated. For the inflatable antenna development, as an extension to the 1-m X-band inflatable reflectarray mentioned above, a 3-m Ka-band circularly polarized inflatable reflectarray consisting of 200,000 elements was also developed [32], which is currently known as the electrically largest reflectarray in the microwave and millimeter-wave spectra. All elements were printed on a thin membrane substrate that was mechanically supported and tensioned at its perimeter by inflatable tubes to achieve a surface rms error of only 0.2 mm. When deflated, the entire antenna can be rolled up to form a 3-m x 0.5-m cylindrical structure for launch vehicle stowage. Another significant development was an amplifying reflectarray [33], as shown in Fig. 2.9. In this reflectarry, each element receives the signal from the feed, the signal then goes through an amplifier, and is then retransmitted, thus achieving a very high overall equivalent isotropic radiated power (EIRP).

Another excellent contribution was the use of the Genetic Algorithm (GA) [34] to optimize the reflectarray element and, thus, to improve the overall antenna efficiency. In a reflectarray, the most critical component is its element, which has many design parameters, such as the substrate permittivity, substrate thickness, patch size, incident angle, main beam direction, bandwidth, etc. The

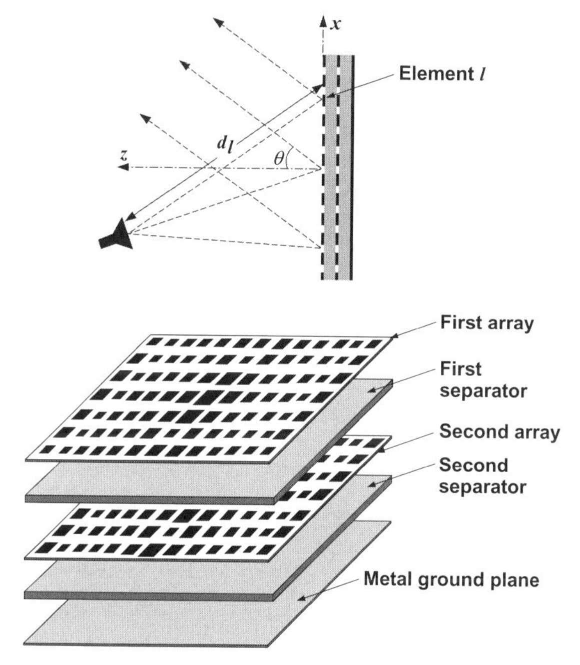

Figure 2.8. Multilayer reflectarray to achieve wider bandwidth.

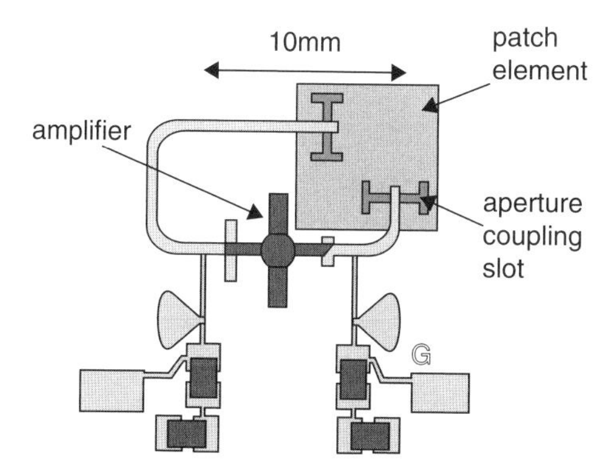

Figure 2.9. Unit-cell element of an X-band amplifying reflectarray.

reflectarray performance in terms of antenna gain and sidelobe level can be optimized with all these element parameters considered in the design by using the GA method. GA is an optimization methodology that allows for the thorough search of a cost surface (reflectarray performance) through stochastic processes that are modeled on the life evolution concept with selection and genetic processes (element parameters).

The reflectarray using subreflector and array offset feed in a Cassegrain configuration to achieve fine beam scanning was also studied with hardware demonstration [35, 36]. In addition to the ability of fine beam scanning, the Cassegrain offset-feed system also reduces reflectarray profile height, as well as minimizes the feed blockage effect.

To combat the shortcoming of narrow bandwidth, dual-band multilayer reflectarrays using annular rings [37], stacked patches [38], and crossed dipoles [39] have also been developed. When two widely separated frequencies, such as X-band and Ka-band, are needed, a single-layer reflectarray will not have enough bandwidth to cover both frequencies. Consequently, a dual-layer structure was developed, where the top-layer elements behave nearly transparent to the bottom-layer signal.

Another innovative technology is the combining of the reflectarray elements with solar cells to form a single panel structure [40]. By doing so, the two largest structures of a spacecraft could be combined to significantly reduce the required real estate. Fig. 2.10 shows a 0.5-m experimental model where thin crossed dipoles are used as reflectarray elements, which, as indicated from measured results, did not impact the solar cell performance.

One other innovative development that is worth mentioning here is the folded reflectarray configuration [41, 42], where two reflecting surfaces, as depicted in Fig. 2.11, are used to significantly reduce the overall antenna profile due the feed height of a conventional reflectarray. In this configuration, the feed horn radiates a linearly polarized wave that is reflected back by the polarizing grid and then illuminates all the reflectarray elements. Each element, upon receiving the incoming wave, not only adjusts the required compensating

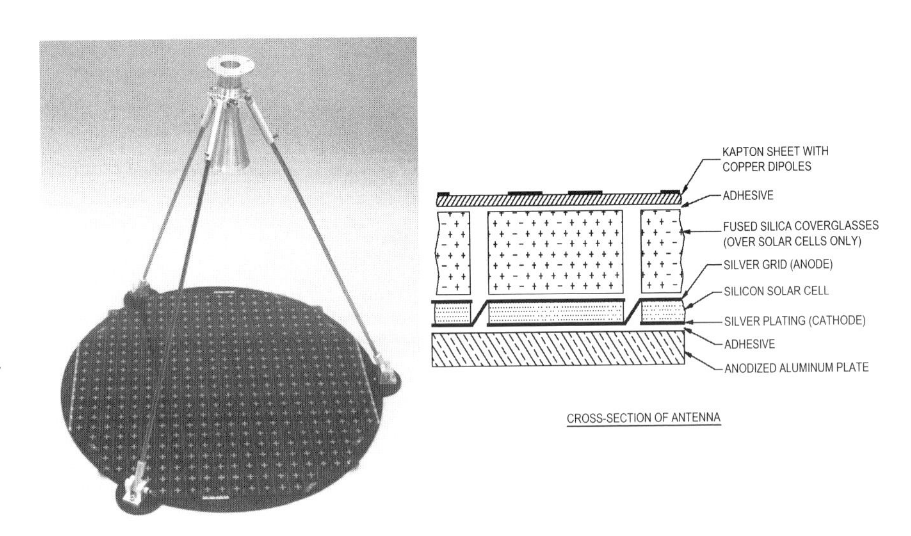

Figure 2.10 Reflectarray elements combined with solar cells.

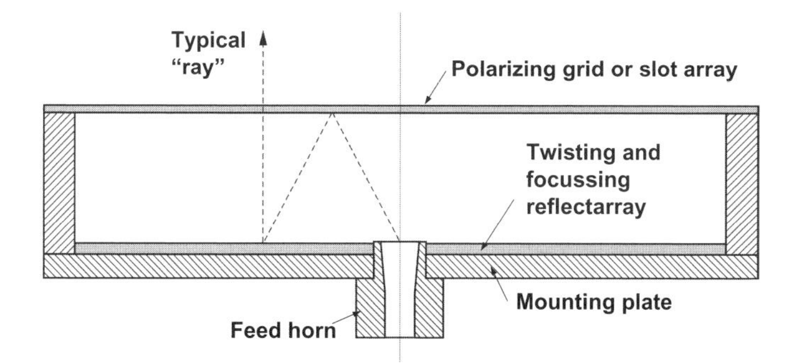

Figure 2.11. Folded reflectarray to achieve more compact profile.

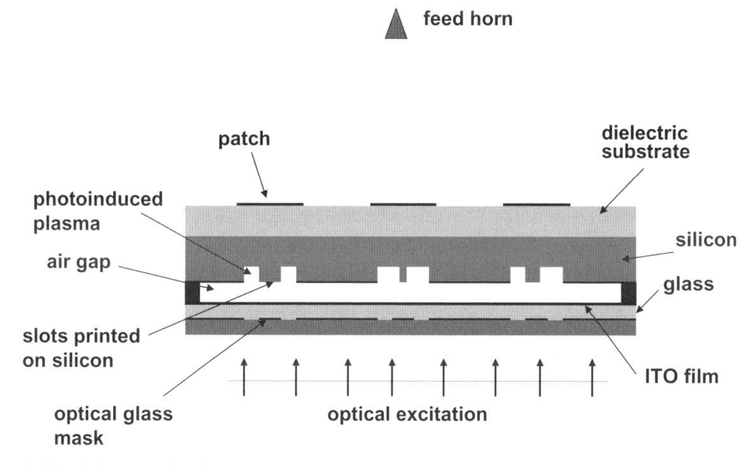

Figure 2.12. Schematic view of the photonically controlled reflectarray. (From [43], © April 2006 IEEE.)

phases but also transforms the linear polarized wave to orthogonal polarization and reradiates through the polarizing grid into far–field space.

The most recent significant innovation was a photonically controlled reflectarray [43] where the elements of a microstrip reflectarray are excited by aperture coupling slots whose lengths are made tunable by photo-induced plasma. The antenna, shown in Fig. 2.12, operates at 30 GHz and is comprised solely of a silicon wafer as the substrate layer. The patch elements of identical size are etched on the top surface of silicon. The bottom surface consists of an array of slots of identical sizes that are centered beneath the patches. The pumping light illuminates the slots through an optical mask and an indium tin-oxide (ITO) film, creating a plasma pattern of the same shape as the pattern on the mask. The ITO film, which is optically transparent, serves as an RF ground plane for the patches. The amount of plasma profile being illuminated by the optical light changes the slot length, which modifies the phase distribution of the reflectarray elements and, hence, allows formation of a co-phasal beam in the desired direction. The plasma profile distribution is controllable to reconfigure the RF beam pattern. Dynamic control of this distribution remains to be developed. This technology, if successfully developed, could significantly reduce the control circuit complexity of controlling many thousands of reflectarray elements.

Finally, another breakthrough that is worth mentioning here is a recently developed patch reflectarray at the infrared frequency with a wavelength of 10.6 μm [44]. It used electronic-beam lithography techniques to fabricate the antenna with 17.9 million patch elements.

2.3 COMPARISON WITH SIMILAR TECHNOLOGIES

There are two antenna technologies that are very similar to the reflectarray system. These are the array lens and the Fresnel-Zone plate reflector. Each technology, when compared with the reflectarray, has its advantages as well as disadvantages. These two technologies are separately discussed below.

2.3.1 Array Lens

An array lens [45, 46], as shown in Fig. 2.13, consists of a feed and an array of isolated elements in a similar fashion as the reflectarray. These elements can be low-profile microstrip patches located either on a planar surface or on a

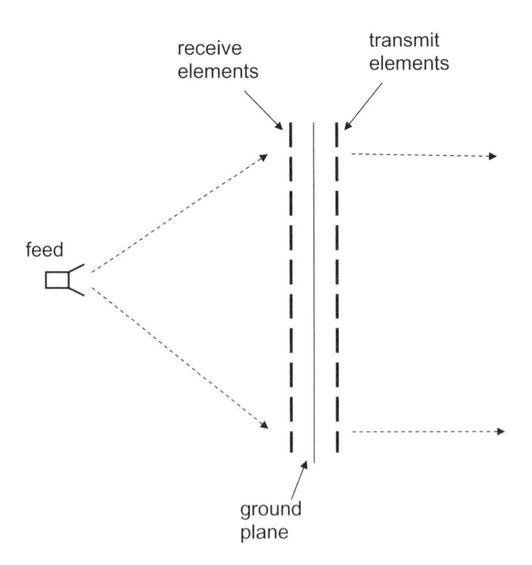

Figure 2.13. Configuration of an array lens.

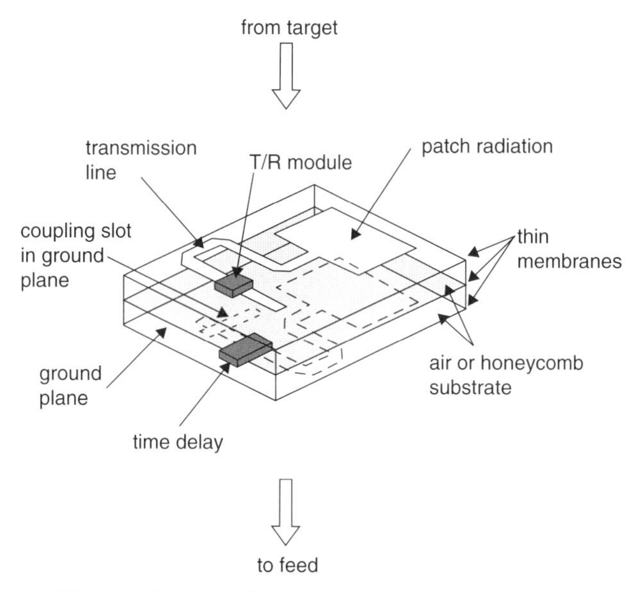

Figure 2.14. A typical element cell in an array lens.

slightly curved surface. These elements, unlike the reflectarray elements where they are reflective, are transmission type. Each element [47], as illustrated in Fig. 2.14, generally consists of two microstrip patches with one receiving energy from the feed and the other transmitting the energy into space on the other side of the lens. Between the receive and transmit elements, there is a short transmission line whose length can be designed to compensate for the path length difference (from a reference element) from the feed. Electronically controlled phase shifters and/or T/R amplifier modules can be placed in these transmission lines to achieve beam scanning. As can be seen, the array lens is very similar in principle, design and operation to the reflectarray.

There are, however, some differences between the array lens and the reflectarray. In terms of physical construction, one primary difference is that the reflectarray is relatively simpler and can be considered almost as half of the array lens. The array lens encompasses both an incoming volume space from the feed and an outgoing volume space for its radiation; while in a reflectarray system, both the incoming volume and the reflection volume share the same space. As an example, a reflectarray can be easily mounted on the sidewall of a building with its small feed supported outside the wall, while an array lens would be difficult to be mounted on the sidewall without breaking into the building to place its feed. In addition, an array lens requires both sets of receive and transmit elements, while in a reflectarray only one set of reflective elements is needed.

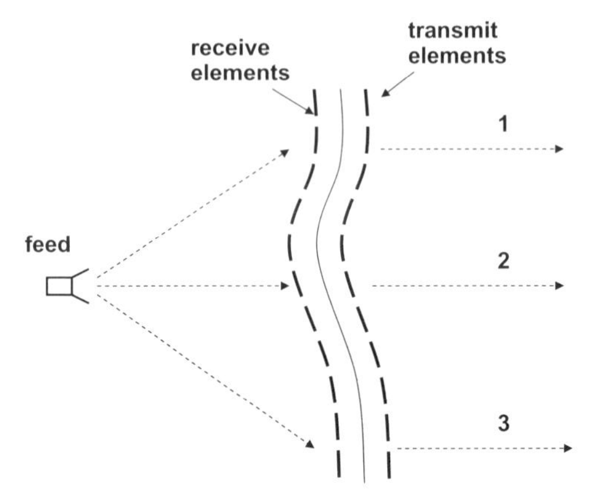

Figure 2.15. Array lens showing insensitivity of path lengths 1, 2, 3 to array surface deviation.

The array lens does have one significant advantage over the reflectarray. When the surface tolerance becomes an issue, for example, a large aperture at Ka-band, the array lens is much more tolerant of surface error than the reflectarray. This is true because the lens transmits energy through and, thus, the path length from the feed to far-field distance does not change due to a surface deviation as shown in Fig. 2.15. In a reflectarray, similar to a parabolic reflector, due to the reflection mechanism, the path length and, thus, the electrical phase can be greatly impacted by a surface deviation.

2.3.2 Fresnel-Zone Plate Reflector

The Fresnel-Zone plate reflector (FZPR) [48, 49] was developed many years ago as an inexpensive alternative to a parabolic reflector. It consists of a feed and a reflecting flat plate. On this reflecting plate, as shown in Fig. 2.16, there are many concentric zones, which are separated by two different heights or thicknesses. The difference between these two heights is $^1/_4$ free-space wavelength (λ_o). The width of each zone (or distance from one ring to the next ring) varies from zone to zone. It is determined by the path length difference between the feed to one zone and the feed to the next zone. This path length difference should be $^1/_2\ \lambda_o$. Because the incident waves from the feed encounter a height difference of $^1/_4\ \lambda_o$ when reaching two adjacent zones and due to the round trip reflection, all reflected waves are more-or-less in phase.

It should be clear by now why the different zones have different widths and why the zone width reduces as it moves away from the center for a center-fed FZPR. Depending on the focal length design, the zone close to the center could have a width of several wavelengths. Instead of the two different heights, a Fresnel-Zone reflector can also be made of a single layer by printing differ-

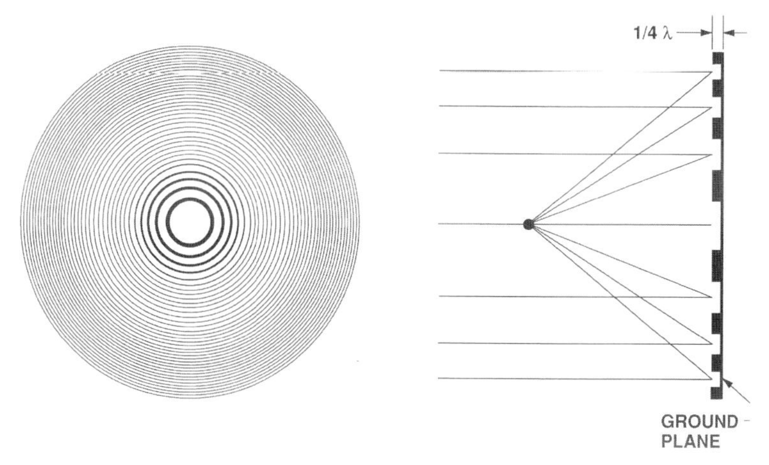

Figure 2.16. Fresnel Zone plate reflector.

ent zones with different-sized elements [18] and thus different sets of reflecting phases as that shown previously in Fig. 2.6.

There is one principal drawback of the FZPR when compared with the reflectarray. The FZPR has relatively low efficiency [50] due to the phase dispersion of the reflected signals in each zone. In other words, not all reflected signals from a "flat" zone are precisely in phase. This inefficiency could result in several decibels loss in antenna gain. Another drawback of the FZPR, similar to the reflectarray, is its narrow bandwidth characteristics. This is because, due to the nature of its flat surface and all waves from the feed to the flat surface have different path lengths, a differential spatial phase delay will occur between waves as frequency changes, which limits the bandwidth performance. This bandwidth limiting effect of the reflectarray or FZPR will be discussed in more detail in a later chapter. Due to the lack of individual elements, the FZPR by itself cannot achieve beam scanning, which is another shortcoming when compared with a reflectarray.

In spite of the above drawbacks, the FZPR does have one distinct advantage. This is the simplicity in designing and manufacturing of the antenna. One should be able to design a FZPR once one understands the basic principle briefly presented above. No lengthy equations or computer are needed. At the frequency around 1 GHz or lower, the antenna can be made in a home garage with plywood and aluminum foil or metallic spray paint since the surface tolerance is not an issue at the low frequencies. At the other end of the frequency spectrum, for example, millimeter-wave frequencies, a FZPR can be easily and accurately made from an aluminum plate with a numerically controlled milling machine without the elaborate step of making a curved mold for a parabolic reflector or the precision chemical etching process for a reflectarray.

REFERENCES

1. D. G. Berry, R. G. Malech, and W. A. Kennedy, "The reflectarray antenna," *IEEE Trans. Antennas Propagat.*, Vol. AP-11, Nov. 1963, pp. 645–651.
2. H. R. Phelan, "Spiralphase reflectarray for multitarget radar," *Microwave Journal*, Vol. 20, July 1977, pp. 67–73.
3. C. S. Malagisi, "Microstrip disc element reflect array," Electronics and Aerospace Systems Convention, Sept. 1978, pp. 186–192.
4. J. P. Montgomery, "A microstrip reflectarray antenna element," Antenna Applications Symposium, University of Illinois, Sept. 1978.
5. R. E. Munson and H. Haddad, "Microstrip reflectarray for satellite communication and RCS enhancement and reduction," U.S. patent 4,684,952, Washington, D.C., August 1987.
6. J. Huang, "Microstrip reflectarray," IEEE AP-S/URSI symposium, London, Canada, June 1991, pp. 612–615.
7. T. A. Metzler, "Design and analysis of a microstrip reflectarray," Ph.D. Dissertation, University of Massachusetts, September 1992.
8. Y. Zhang, K. L. Wu, C. Wu, and J. Litva, "Microstrip reflectarray: full-wave analysis and design scheme," IEEE AP-S/URSI symposium, Ann Arbor, Michigan, June 1993, pp. 1386–1389.
9. R. D. Javor, X. D. Wu, and K. Chang, "Beam steering of a microstrip flat reflectarray antenna," IEEE AP-S/URSI symposium, Seattle, Washington, June 1994, pp. 956–959.
10. D. C. Chang and M. C. Huang, "Multiple polarization microstrip reflectarray antenna with high efficiency and low cross-polarization," *IEEE Trans. Antennas Propagat.*, Vol. 43, August 1995, pp. 829–834.
11. A. Kelkar, "FLAPS: conformal phased reflecting surfaces," Proc. IEEE National Radar Conf., Los Angeles, California, March 1991, pp. 58–62.
12. D. M. Pozar and T. A. Metzler, "Analysis of a reflectarray antenna using microstrip patches of variable size," *Electronics Letters*, April 1993, pp. 657–658.
13. D. M. Pozar, S. D. Targonski, and H. D. Syrigos, "Design of millimeter wave microstrip reflectarrays," *IEEE Trans. Antennas Propagat.*, Vol. 45, Feb. 1997, pp. 287–296.
14. K. Y. Sze and L. Shafai, "Analysis of phase variation due to varying patch length in a microstrip reflectarrays," IEEE AP-S/URSI symposium, June 1998, pp. 1134–1137.
15. J. Huang, "Bandwidth study of microstrip reflectarray and a novel phased reflectarray concept," IEEE AP-S/URSI symposium, Newport Beach, California, June, 1995, pp. 582–585.
16. J. Huang and R. J. Pogorzelski, "A Ka-band microstrip reflectarray with elements having variable rotation angles," *IEEE Trans. Antennas Propagat.*, Vol. 46, May 1998, pp. 650–656.
17. F. S. Johansson, "A new planar grating-reflector antenna," *IEEE Trans. Antennas Propagat.*, Vol. 38, Sept. 1990, pp. 1491–1495.
18. Y. T. Gao and S. K. Barton, "Phase correcting zonal reflector incorporating rings," *IEEE Trans. Antennas Propagat.*, Vol. 43, April 1995, pp. 350–355.

19. J. M. Colin, "Phased array radars in France: present and future," IEEE symposium on Phased Array System and Technology, Boston, Massachusetts, October 1996, pp. 458–462.
20. A. A. Tolkachev, V. V. Denisenko, A. V. Shishlov, and A. G. Shubov, "High-gain antenna system for millimeter-wave radars with combined electrical and mechanical beam steering," IEEE symposium on Phased Array System and Technology, Boston, Massachusetts, October 1996, pp. 266–271.
21. L. Boccia, F. Venneri, G. Amendola, and G. Di Massa, "Application of varactor diodes for reflectarray phase control," IEEE AP-S/URSI Symposium, San Antonio, Texas, June 2002, Vol. 3, pp. 132–135.
22. S. V. Hum and M. Okoniewski, "An electronically tunable reflectarray using varactor-diode-tuned elements," IEEE AP-S/URSI Symposium, Monterey, California, June 2004, Vol. 2, pp. 1827–1830.
23. S. V. Hum, G. McFeetors, and M. Okoniewski, "A reflectarray cell based on a tunable MEMS capacitor," IEEE AP-S/URSI Symposium, Albuquerque, New Mexico, July 2006, URSI session 458.
24. A. Martynyuk, J. Lopez, J. Cuevas, and Y. Sydoruk, "Wideband reflective array based on loaded metal rings," IEEE MTT-S Microwave symposium, Long Beach, California, June 2005.
25. V. F. Fusco, "Mechanical beam scanning reflectarray," *IEEE Trans. Antennas Propag.*, Vol. 53, Nov. 2005, pp. 3842–3844.
26. R. H. Phillion and M. Okoniewski, "Measuring the reflection from a polarization selective reflectarray element," IEEE AP-S/URSI Symposium, Albuquerque, New Mexico, July 2006, pp. 4343–4347.
27. J. Huang and A. Feria, "A 1-m X-band inflatable reflectarray antenna," *Microwave and Optical Technology Letters*, Vol. 20, January 1999, pp.97–99.
28. J. Huang, "Capabilities of printed reflectarray antennas," IEEE symposium on Phased Array System and Technology, Boston, Massachusetts, October 1996, pp. 131–134.
29. D. M. Pozar, S. D. Targonski, and R. Pokuls, "A shaped-beam microstrip patch reflectarray," *IEEE Trans. Antennas Propagat.*, Vol. 47, July 1999, pp. 1167–1173.
30. J. A. Encinar, "Design of two-layer printed reflectarray using patches of variable size," *IEEE Trans. Antennas Propagat.*, Vol. 49, October 2001, pp. 1403–1410.
31. J. A. Encinar and J. A. Zornoza, "Broadband design of three-layer printed reflectarrays," *IEEE Trans. Antennas Propagat.*, Vol. 51, July 2003, pp. 1662–1664.
32. J. Huang, V. A. Feria, and H. Fang, "Improvement of the three-meter Ka-band inflatable reflectarray antenna," IEEE AP-S/URSI symposium, Boston, Massachusetts, July 2001, pp. 122–125.
33. M. Bialkowski, A. W. Robinson, and H. J. Song, "Design, development, and testing of X-band amplifying reflectarrays," *IEEE Trans. Antennas Propagat.*, Vol. 50, August 2002, pp. 1065–1076.
34. R. E. Zich, M. Mussetta, M. Tovaglieri, P. Pirinoli, and M. Orefice, "Genetic Optimization of microstrip reflectarrays," IEEE AP-S/URSI symposium, San Antonio, Texas, June 2002, pp. III-128–131.
35. B. Khayatian and Y. Rahmat-Samii, "Characterizing reflectarray antenna radiation performance," IEEE AP-S/URSI symposium, Columbus, Ohio, June 2003.

36. J. Huang, C. Han, and K. Chang, "A Cassegrain offset-fed dual-band reflectarray," IEEE AP-S/URSI symposium, Albuquerque, New Mexico, July 2006, pp. 2439–2442.
37. C. Han, C. Rodenbeck, J. Huang, and K. Chang, "A C/Ka dual-frequency dual-layer circularly polarized reflectarray antenna with microstrip ring elements," *IEEE Trans. Antennas Propagat.*, Vol. 52, Nov. 2004, pp. 2871–2876.
38. J. A. Encinar, "Design of a dual-frequency reflectarray using microstrip stacked patches of variable size," *Electronic Letters*, Vol. 32, No. 12, June 1996. pp. 1049–1050.
39. M. Zawadzki and J. Huang, "A dual-band reflectarray for X- and Ka-bands," PIERS symposium, Honolulu, Hawaii, October 2003.
40. M. Zawadzki and J. Huang, "Integrated RF antenna and solar array for spacecraft application," IEEE Phased Array System and Technology Conference, Dana Point, California, May 2000, pp. 239–242.
41. W. Menzel, D. Pilz, and M. Al-Tikriti, "Millimeter-wave folded reflector antennas with high gain, low-loss, and low profile," feature article, *IEEE Antennas & Propagation Magazine*, Vol. 44, No. 3, June 2002, pp. 24–29.
42. J. A. Zornoza, R. Leberer, J. A. Encinar, and W. Menzel, "Folded Multilayer microstrip reflectarray with shaped pattern," *IEEE Trans. Antennas Propagat.*, Vol. 54, February 2006, pp. 510–518.
43. M. R. Chaharmir, J. Shaker, M. Cuhaci, and A. R. Sebak, "Novel photonically-controlled reflectarray antenna," *IEEE Trans. Antennas Propagat.*, Vol. 54, April 2006, pp. 1134–1141.
44. J. C. Ginn, B. A. Lail, and G. D. Boreman, "Infrared patch reflectarray," IEEE AP-S/URSI symposium, Albuquerque, New Mexcio, July 2006, pp. 4315–4318.
45. R. Tang, "Survey of time-delay steering techniques," Phased Array Antennas: Proc. 1970 Phased Array Antenna Symp., Dedham, Massachusetts, Artech House, 1970, pp. 254–260.
46. S. Dattanasombat, A. Prata, L. R. Amaro, and J. A. Harrel, "Layered lens antennas," IEEE AP-S/URSI symposium, Boston, Massachusetts, July 2001, pp. 777–780.
47. D. T. McGrath, P. M. Proudffot, and M. A. Mehalic, "The microstrip constrained lens," Microwave Journal, January 1995, pp. 24–37.
48. L. F. Van Buskirk and C. E. Hendrix, "The zone plate as a radio frequency focusing element," *IRE Trans. Antennas Propagat.*, Vol. AP-9, May 1961, pp. 319–320.
49. J. C. Wiltse and J. E. Garrett, "The Fresnel zone plate antenna," *Microwave Journal*, Vol. 34, January 1991, pp. 101–114.
50. J. Gutierrez-Rios and J. Vassal'lo, "Technology Aspects of Fresnel Zone reflectors," Chapter 4 in *Advances on Antennas, Reflectors and Beam Control*, edited by A. Tazon, Research Signpost, Kerala, India, 2005.

CHAPTER 3

Antenna Analysis Techniques

3.1 INTRODUCTION

A printed reflectarray consists of a planar array of printed radiating elements that incorporates a certain phase-shift to produce a collimated or a shaped beam when it is illuminated by a feed (Fig. 3.1). The operating principle can be explained by considering the reflectarray in transmitting mode with a horn antenna located in a centered or offset position, and assuming that the reflectarray elements are in the far-field region of the horn. In this case, the electromagnetic field incident on each reflectarray element at a certain angle can be locally considered as a plane wave with a phase proportional to the distance from the phase center of the feed-horn to each element, as corresponds to a spherical wave propagation. In order to convert the spherical wave radiated by the horn into a focused beam, the field must be reflected from each element with an appropriate phase shift. This phase shift is adjusted independently for each element to produce a progressive phase distribution of the reflected field on the planar surface that generates a pencil beam in a given direction, as is known from classic phased array theory. Alternatively, contoured beams can be generated by implementing the appropriate phase distribution, obtained by a phase-only synthesis method.

The necessary phase shift at each element is obtained by varying one of the geometrical parameters in the reflectarray element. The first implementation of phase adjustment in rectangular microstrip patches consists of connecting transmission line segments of different lengths to the printed elements [1–4] (Fig. 3.2(a)). In this technique, the phase delay of the reflected wave is proportional to the length of the stubs. However, the stubs produce some dissipative losses and spurious radiation when bent. Other concepts for microstrip reflectarrays have been developed, such as the adjustment of resonant length in dipoles, crossed dipoles, or rectangular patches [5–8], aperture-coupled patches with stubs [9, 10], and apertures of different length on a metal plane [11].

Reflectarray Antennas, by John Huang and José A. Encinar

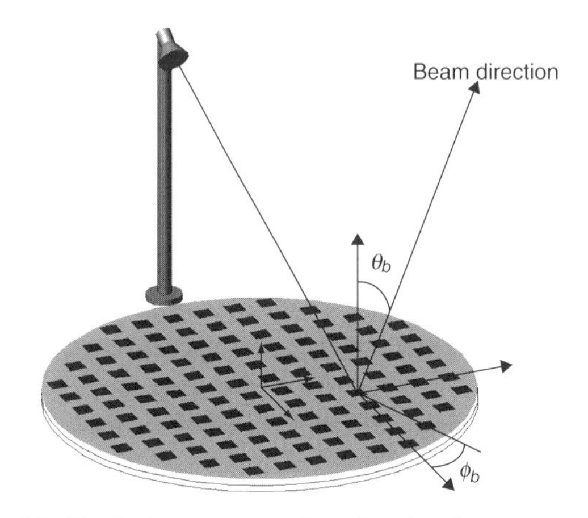

Figure 3.1. Typical geometry of a printed reflectarray antenna.

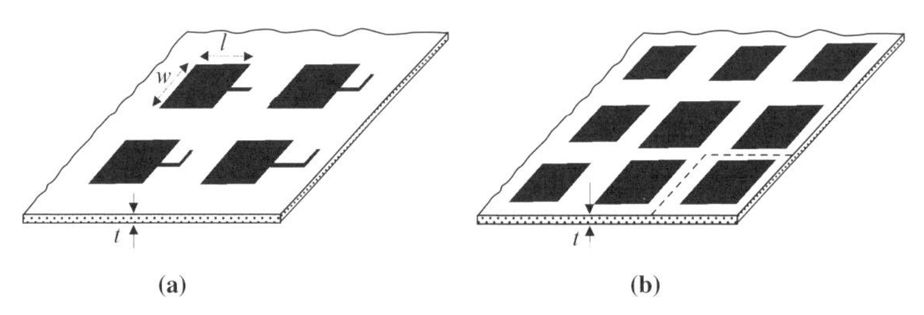

Figure 3.2. Phasing elements in printed reflectarrays. (a) Rectangular patches with attached stubs. (b) Rectangular patches of varying size.

In principle, these implementations are valid for any type of polarization, including dual-linear or circular polarization, by appropriately adjusting the phase-shift of the two orthogonal components of the reflected electric field. For circular polarization, other concepts have been demonstrated, such as elements with variable rotation angles [12], and spiraphase-type reflectarrays based on loaded ring slot resonators [13].

The concept of phase control by varying the resonant dimensions, as shown in Fig. 3.2(b), produces lower dissipative losses and cross-polarization levels than the stubs of different lengths attached to the radiating patches [5–8]. However, the maximum range of phase variation that can be achieved is on the order of 330°, and the phase variation versus the length is strongly nonlinear because of the narrow band behavior of microstrip patches. In addition, the phase is very sensitive to variations in frequency, which limits the operating bandwidth of the reflectarray. All these drawbacks are solved by using two or three stacked array layers [14–16].

An alternative reflectarray element is based on aperture-coupled patches with stubs of different lengths [9, 10]. In this configuration, as in that of the attached-stub concept, a real phase delay proportional to the stub length is achieved. The aperture-coupled element has several advantages, as is known from printed arrays: 1) there is more room for longer lines in order to increase the range of phase delay, 2) the spurious radiation produced by the stubs is in the opposite direction than the antenna beam, and 3) active elements or controllable phase-shifters can be included in the microstrip lines to reconfigure or scan the beam.

For the design of the reflectarray, any possible value of phase-shift must be implemented by varying one parameter in the unit cell, such as the patch size, stub length, or patch rotation angle. One of the most important parts of the reflectarray analysis and design is the accurate characterization of the reflective elements, that is, for a given geometry of the reflectarray element, to accurately predict the phase-shift and dissipative losses for each polarization of the field. These data can be obtained from simple circuit models [3, 4], or from computer simulations based on a full-wave electromagnetic analysis technique [6, 17, 18], as discussed later.

Once the unit element has been fully characterized, an important aspect of the reflectarray analysis is the accurate evaluation of the radiation patterns, including co- and cross-polar components. For the evaluation of the radiation patterns, the feed model has to be taken into account. Typically, the feed-horn is modeled as a $\cos^q(\theta)$ function, but more accurate simulations or measurements can be used in the evaluation of the radiation patterns of the horn. The radiation patterns can be computed in gain if the power patterns are divided by the total power in the feed-horn. These points are addressed in the final section of this chapter.

3.2 OVERVIEW OF ANALYSIS TECHNIQUES

An extended approach for the analysis and design of reflectarray antennas is based on the use of design curves relating the phase of the reradiated field with a certain geometrical parameter of the reflectarray element assuming normal incidence. This approach assumes that the phase response is independent of the angle of incidence of the impinging wave. This assumption, usually adopted, is valid for the central elements of the centered fed reflectarray, that is for angles not too far from normal incidence. As the largest fraction of the incident power is reflected by the central part of the reflectarray, the normal incidence assumption can provide good predictions for center-fed reflectarrays with reasonably large F/D. It was shown in reference [7], that for a 40° angle of incidence on rectangular patches of variable size, the phase changes around 25° with respect to the response of normal incidence. The change in phase goes to 50° for an incidence angle of 60°. Note that the offset configuration is normally used to avoid the feed shadow effect and the angles of incidence are on the order of 40° and larger near one reflectarray edge. Thus, this approach may not be accurate for those elements near the edge, and the angle of incidence must be considered in the design.

An important aspect of the analysis of a reflectarray is the accurate prediction of the cross-polarization. Although it is assumed that reflectarrays produce lower cross-polarization than classic offset reflector antennas [6, 7], it must be evaluated. The cross-polarization is mainly generated by the bent stubs in the elements with attached stubs, but it is also generated by the rectangular patches, or the corresponding reflective element, when the angle of incidence is out of the principal planes. Note that, under the assumption of normal incidence, the cross-polarization cannot be predicted. In order to compute the cross-polarization generated by each reflectarray element, the incident field coming from the feed at an angle of incidence (θ, φ) must be broken down into two orthogonal components, parallel to the rectangular lattice of the reflectarray. Then, the two components of the field reflected on each element must be computed using a full-wave analysis method. However, the real angles of incidence (θ, φ) for each element cannot be taken into account using design curves, simply because of the very large number of combinations. Therefore, an alternative method should be used for a more accurate analysis of each element in order to evaluate the cross-polarization generated by the reflectarray elements.

When analyzing the reflectarray elements using a full-wave technique, it must be taken into account that the radiated field is the sum of two components: the field reradiated by the patches and the field reflected from the ground plane. The second component, which is not present in array antennas, must be included in the analysis method, for example, in the Method of Moments. It was demonstrated in [19] that, when this contribution is not considered, a completely different result in phase can be obtained from the simulations.

The reflective element can be analyzed as an isolated element or in an array environment. One approach used for the analysis and design of reflectarrays is based on the assumption that each element behaves as an isolated element on a grounded dielectric substrate [4, 20, 21], and mutual coupling between radiating elements is neglected. This assumption is very common in reflectarrays with attached stubs [4, 20], but has also been used for variable size patches [21]. It was stated in [4] that mutual coupling can be neglected when the distance between the edges of adjacent patches is larger than 0.25 wavelengths in the dielectric, as already used in printed arrays [22]. The printed elements are normally arranged in periodic square lattice whose sides are around 0.6–0.7 wavelengths, in order to avoid the appearance of grating lobes for any angle of incidence. Then, the distance between patch edges can be smaller than 0.25 λ and mutual coupling may not be negligible. For example, the phase-shift computed for a 23-element linear array of stub-loaded patches periodically distributed, with half-a-wavelength period, shows differences of more than 50° with respect to the phase-shift computed for an isolated element [17, 18]. The effect of mutual coupling can be stronger for reflectarrays based on variable size patches, because the dimensions of some patches are larger than half-wavelength in the dielectric and the separation between patches is small.

A simple method to account for mutual coupling effects was proposed in [23] by using the Finite Difference Time Domain (FDTD) modeling technique. It consists of illuminating a single cell in the array in the presence of the nearest neighbor cells (identical or not) and calculating the reflected wave. This technique allows the reflection coefficient of each element to be computed taking into account the real mutual coupling of the neighboring elements, but it is impractical for the design of a large reflectarray.

For the analysis of arrays with a large number of elements, when element-by-element mutual coupling analysis techniques are prohibitive, the infinite array model is usually used and, by applying Floquet's theorem, the analysis is reduced to only one periodic cell [24, 25]. This technique automatically takes into account the mutual coupling between elements and provides good predictions of each element in the array environment. The analysis of the reflectarray element using the infinite array approach is very efficient, as well as being accurate in characterizing the reflectarray elements, as reported in [19].

Different numerical methods have been implemented for a full-wave analysis of periodic structures, such as Method of Moments in the Spectral Domain for multilayer configurations [26, 27], Finite Element Method (FEM) [28], or FDTD [29]. Zero thickness is usually assumed for the conductors in the analysis methods based on integral equations solved by the Method of Moments (MoM). On the other hand, FEM and FDTD allow the real conductor thickness in the simulations to be taken into account. For example, a hybrid technique based on Generalized Scattering Matrix and FEM was used to analyze an array of crossed-slots in a conductor slab of finite thickness [30]. The FDTD technique has also been applied to analyze reflectarray cells of complex geometry [31, 32].

A unit cell for circular polarization, made of a circular conductive patch with several radial stubs that can be connected or disconnected, was analyzed in [31] and a unit cell based on a dielectric resonator was analyzed in [32]. The analysis in time domain allows the response in a wide frequency band to be computed, but it is more time consuming than the Method of Moments.

Spectral Domain Method of Moments [6, 7] is very appropriate and numerically efficient for a full-wave analysis of periodical structures, assuming planar arrays of patches or apertures in a single- or multilayer configuration. In this case, the CPU time for the analysis is very low, and the analysis routine can be integrated into an optimization loop for a more accurate design of reflectarray antennas. The infinite array approach can be used in principle for any type of phase-shifter element.

For elements with stubs of different lengths, all the radiating patches are exactly the same and only the stub length varies from one element to the next. In this case, the infinite array approach will be very accurate, because the coupling produced by the stubs is less significant, assuming a minimum of separation between the stubs and other metallic lines or patches.

For elements with variable-sized patches, the reflectarray is analyzed by assuming local periodicity, that is, each element is considered in an array environment with all the elements identical. This approach is accurate when the variation in patch dimensions is smooth from one cell to the next, because it takes into account all mutual coupling between patches. The local periodicity approach is not accurate only in the case of a small number of patches where the surrounding patches are of very different dimensions. Normally, these elements are a small percentage of the total number of elements and the overall radiation pattern of the reflectarray is only slightly affected, as shown in [33].

For the analysis by MoM of noncanonical patches in a periodic environment, the unknown current is usually represented by subdomain basis functions, meaning an increase in CPU time. Alternatively, a set of entire-domain basis functions can be determined for arbitrarily shaped patches by using the boundary integral-resonant mode expansion (BI-RME) method as described in [34, 35]. The resulting basis functions are used in the MoM, and this permits the analysis of noncanonical patches with similar CPU times as for rectangular patches.

As an alternative to the local periodicity approach, different techniques have been proposed to analyze a single layer reflectarray with variable-sized patches [36–38], taking into account the real mutual coupling between elements. A full-wave spectral domain Method of Moments was implemented [36] for the analysis of a reflectarray that radiates a pencil beam, and a good agreement was obtained between measured and predicted radiation patterns.

Two fast computational techniques were presented in [37] for the analyses of microstrip reflectarrays. The first one is the Sparse-Matrix Canonical Grid (SMCG) method, which has a computational complexity of $O(N \log N)$. It allows the use of Fast Fourier Transforms (FFTs) to compute a major portion of the matrix-vector multiplication needed in the iterative solution in the MoM matrix equation. In contrast, the second method adopts a full-matrix inversion scheme with the number of unknowns substantially reduced through the use of Characteristic Basis Functions (CBFs) derived for each array element. Both methods incorporate all mutual coupling between the array elements and yield radiation patterns that are virtually indistinguishable from each other. The numerical results compared favorably with the experimental data. Although CPU time has been drastically reduced using a parallel computer with eight nodes, the CPU time is still more than 50 minutes for the analysis of a single-layer reflectarray with 1,117 elements. Finally, an adaptive integral method was proposed in [38] for the analysis of a single-layer reflectarray by using MoM. A reflectarray of 100×100 elements was analyzed using an IBM p690 16 processor computer and the resulting CPU time was 15 minutes.

Previous CPU times can be acceptable for the analysis but cannot be included in the optimization routines for the design and optimization of the reflectarray because of the large number of iterations required. In conclusion, after the reflectarray design is completed using the local periodicity approach, or other simplified approaches such as design curves, the reflectarray can be analyzed as a whole using a full-wave technique, to estimate the possible discrepancies.

Some authors have used commercial software for the analysis and design of printed reflectarrays [39, 40]. Some electromagnetic simulators, such as HFSS from Ansoft, incorporate the possibility of implementing periodicity conditions in their recent versions. This software tool based on the Finite Element Method can be used to obtain the curves of phase for a reflectarray element in an infinite array environment. However, general-purpose codes are slow and impractical for the analysis and design of a reflectarray antenna.

The phase curves obtained by a simulation tool can be used to determine the dimensions of the reflectarray elements in an approximate design method, ignoring the cross-polarization and the effect of the angle of incidence. On the other hand, for a more accurate design process, such as the one described in [15], or [16] for optimizations in a frequency band, the design program has to call the analysis routine in order to carry out an optimization of the patch dimensions, until the phase requirements for each polarization and frequency (or frequencies) are fulfilled in each element. In this case, the commercial analysis tools are impractical because of the difficulties in defining the interfaces for the implementation of the analysis routine as a part of an external optimization program.

After reviewing the different approaches used for the analysis of reflectarrays, it is clear that the Method of Moments in the spectral domain assuming local periodicity is the best choice for efficiency and accuracy. This method must take into account the real values of the angle of incidence, polarization, and field radiated by the feed. The very low CPU times allow this method to be used in optimization processes such as improving the bandwidth, as will be discussed in a forthcoming chapter.

Some of the most promising concepts for the phase-shifting in reflectarrays, such as variable-sized stacked patches [14–16] and aperture-coupled patches [9, 10], can be considered as a multilayer structure, with periodic surfaces made of metal patches or apertures in a ground plane separated by dielectric layers. The multilayer periodic structure can be analyzed using the modular approach proposed in [27], which consists of characterizing each array layer by a generalized (or multi-mode) matrix and then analyzing the whole structure by using a cascading process. This approach is very flexible for analyzing diverse geometries, since the matrix that characterizes each layer is computed independently, and it is used as a building block for the analysis of multilayer structures. Using this technique, the individual problems for each array or transition involve very simple two-layer Green's functions. The number of unknowns does not increase with the number of layers. An additional layer only requires a single two-layer problem to be solved plus simple matrix operations for the cascade process. This modular technique has been successfully applied to the analysis of Frequency Selective Surfaces (FSS) [27, 41, 42], printed arrays [43], and reflectarrays [15, 16]. The technique is described in this chapter and it is applied to the analysis of two types of reflectarray elements: aperture-coupled stubs and variable-sized stacked patches. Both types of phase-shifting allow a significant improvement in the element bandwidth and a range in phase delay of several times 360°.

3.3 PHASE-SHIFT DISTRIBUTION

The phase-shift that must be introduced at each element to produce a collimated beam in a given direction is determined in this section. Considering the coordinate system detailed in Fig. 3.1, the progressive phase distribution on the reflectarray surface that produces a beam in the direction (θ_b, φ_b), as known from array theory, is expressed as

$$\phi(x_i, y_i) = -k_0 \sin\theta_b \cos\varphi_b x_i - k_0 \sin\theta_b \operatorname{sen}\varphi_b y_i, \tag{3.1}$$

where k_0 is the propagation constant in vacuum, and (x_i, y_i) the coordinates of element i. On the other hand, the phase of the reflected field at each reflectarray element is equal to the phase of the incident field, as a result of propagation from the feed, plus the phase-shift introduced by each cell, as

$$\phi(x_i, y_i) = -k_0 d_i + \phi_R(x_i, y_i), \tag{3.2}$$

where, $\phi_R(x_i, y_i)$ is the phase of the reflection coefficient, or phase-shift, for element i, d_i is the distance from the phase center of the feed to the cell. From expressions (3.1) and (3.2), the phase-shift required at each element is obtained:

$$\phi_R = k_0(d_i - (x_i \cos\varphi_b + y_i \sin\varphi_b)\sin\theta_b) \tag{3.3}$$

Fig. 3.3 shows the required phase-shift on a circular reflectarray of 30 × 30 elements with the focal point centered that produces a pencil beam in a direction normal to the surface. For the reflectarray design, the phase of the reflection coefficient must be adjusted in each element to match theses phases. The possibility of an independent phase adjustment for each reflectarray printed element can be used to shape the beam. In this case, the phase distribution given in Eq. (3.3) must be substituted by an appropriate phase distribution, obtained by a phase-only synthesis method as will be discussed in a later chapter.

The phase-shift (3.3) is achieved by varying one of the geometrical parameters in the reflectarray elements. The most classic implementation of phase adjustment in rectangular microstrip patches consists of connecting transmission line segments of different lengths to the printed elements.

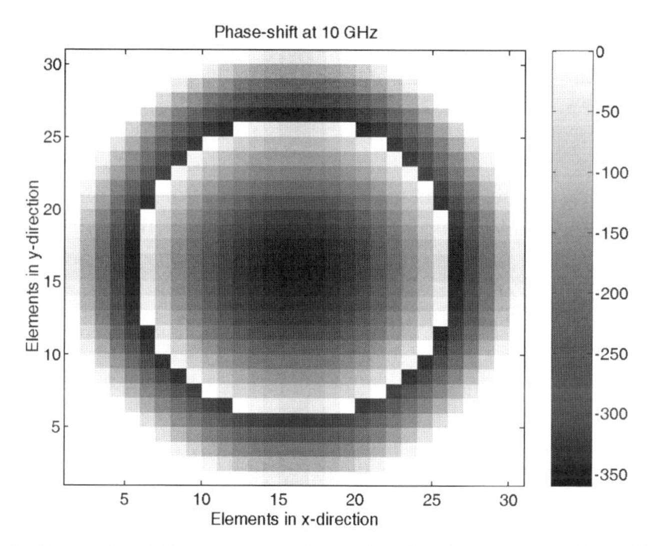

Figure 3.3. Example of the required phase distribution in a circular reflectarray.

3.4 ANALYSIS OF RECTANGULAR PATCHES WITH ATTACHED STUBS

In this configuration, each patch receives the signal from the feed, which is transferred to the stub and propagated until its end, which can be either a short or open circuit, where it is reflected, propagated back, and radiated by the microstrip patch with a phase shift proportional to twice the line length. The analysis and design of this type of reflectarray element is carried out by using the same techniques as in classic microstrip antennas. First, the patch dimensions must be suitably chosen so that it resonates at the frequency of the incident field. Second, the line (stub) must be well matched to the impedance of the radiating patch to allow the transmission of the power to the delay line (stub). The determination of the dimensions of patch and line, as well as point of connection will be carried out by using well-known techniques in the analysis and design of microstrip antennas. Approximate techniques [4, 44], or a commercially available full-wave software can be used for the analysis of the radiating element.

Once the radiating element has been designed, the phase of the reflected field on each reflectarray element is directly controlled by the length of the stub. In reality, the radiated wave exhibits a phase shift with respect to incident wave, which is governed by the length of the open-circuited stub, but it is also affected by the patch itself. However, if two neighboring elements are considered, the angle of incidence is very similar for both of them. The effect of the patch on the phase-shift is also practically the same. Then, the differential phase shift can be estimated as $2\beta\Delta l$, where β is the propagation constant of the signal along the stub and Δl is the physical length difference between the stubs used in the two patches. In conclusion, for analysis and design of the reflectarray, the relative phase-shift can be considered as proportional to twice the length of the line, assuming that patch and stub are well matched.

Note that if a significant mismatch between patch and line occurs, part of the energy is reflected by the patch itself and it is not transmitted to the stub. Then, the total reflected field is the superposition of two components, the one reflected at the patch and the one reflected at the end of the stub. As a result, the phase-shift is no longer proportional to the length of the line. For example, if the patch is initially designed as an isolated element, it may be mismatched to the stub when placed in an array. For a better phase response, the effect of mutual coupling can be taken into account during the design of the reflective element by considering the patch in an array environment.

For the idealized case, the phase response as a function of the stub length must be perfectly linear. However, apart from possible mismatch between the patch and stub, there are other effects not included in the previous simplified analyses approach, that can distort the linear phase response. The first one is the specular reflection on the ground plane that will be added to the energy re-radiated by the patches. The second one is the resonances of the stubs that can happen for certain stub lengths. Both effects can be taken into account in a full-wave analysis of the reflective element [17]. The phase curves versus the stub length computed in [17, 18] for a linear array of 23 elements exhibit a nonlinear behavior produced by the aforementioned effects.

For two-dimensional arrays, the phase-shift versus the stub-length can be computed in a more accurate way by Method of Moments, assuming a periodic array as in [35]. The results shown in [35] for square patches with attached stubs computed by MoM/boundary integral resonant mode expansion (BI-RME) are also very different from the ideal linear expected response. The reflectarray element consists of a rectangular patch w = 10 mm wide and resonant length l = 6.9 mm printed on a grounded dielectric layer t = 1.52 mm thick and ε_r = 3.5 with a stub 0.11 mm wide attached on the center of the radiating side in a periodic cell of 20 × 20 mm, as shown in Fig. 3.2(a).

The phase response has also been computed by SD-MoM, using roof-top basis functions to represent the current distribution on the patch and stub for the same reflectarray element, as described in [27]. The phase values of the reflected field for normal incidence versus the stub-length are shown in Fig. 3.4 at 10 GHz, which are very similar to those reported in [35]. Note that the stub width (0.11 mm) is extremely narrow in comparison with the patch width (10 mm), which requires a very large number of roof-top functions if a uniform rectangular lattice is considered. The results in Fig. 3.4 have been obtained using 450 plus 373 basis-functions for x- and y-current distribution, in a rectangular lattice of 0.11 mm × 1.33 mm, respectively. The grid has been chosen as a compromise between accuracy and numerical efficiency, but requires a slight modification in the patch dimensions (10.05 mm × 6.67 mm), and may

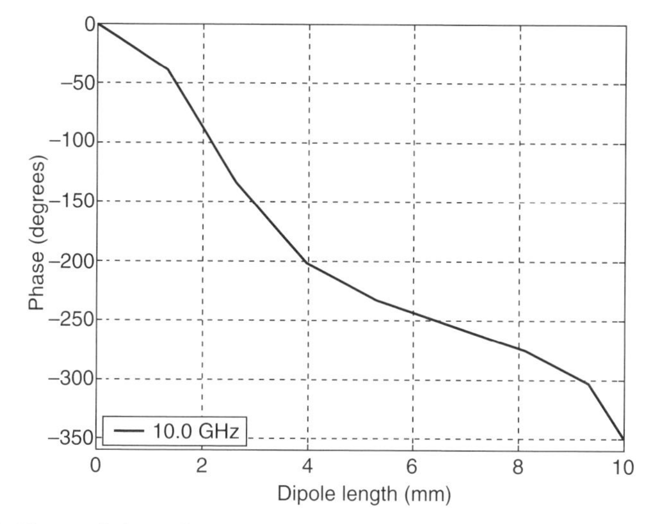

Figure 3.4. Phase of the reflected field for normal incidence versus the stub-length (t = 1.52 mm, ε_r = 3.5).

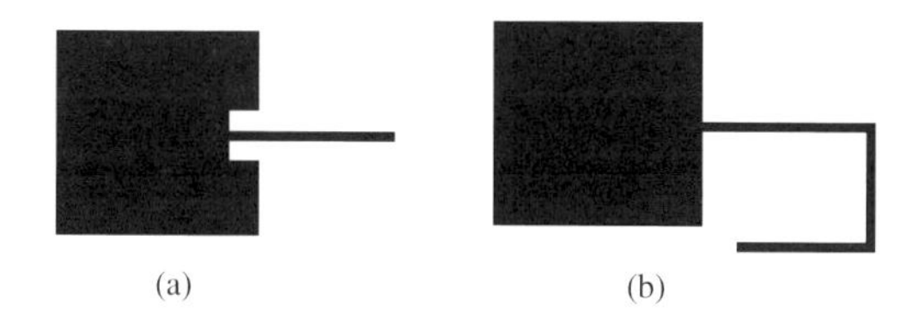

Figure 3.5. Patches with attached stubs. (a) Indented line, (b) folded line.

produce a lack of accuracy in modeling the current distribution in the stubs. These two effects could be the reason for the slight differences between the phases in Fig. 3.4 and in [35].

A similar patch of dimensions w = 10.5 mm and l = 7 mm with attached stub of width 0.7 mm has been analyzed at 12 GHz considering a periodic lattice of $0.5\lambda \times 0.5\lambda$. For the attachment of the stubs, an indentation that consists of connecting the line to the patch at a point placed 1.4 mm within the border has been considered, in order to improve the matching between the line and the patch, see Fig. 3.5(a). The results show that the phase-shift varies with the indentation, but the linearity in phase is not improved as can be seen in Fig. 3.6. The same figure shows that the total range in phase delay can be increased up to −540°, by bending the stubs and increasing their lengths, as shown in Fig. 3.5(b). The ohmic losses in the dielectric layer are computed as the amplitude of the reflection coefficient and are shown in Fig. 3.6(b).

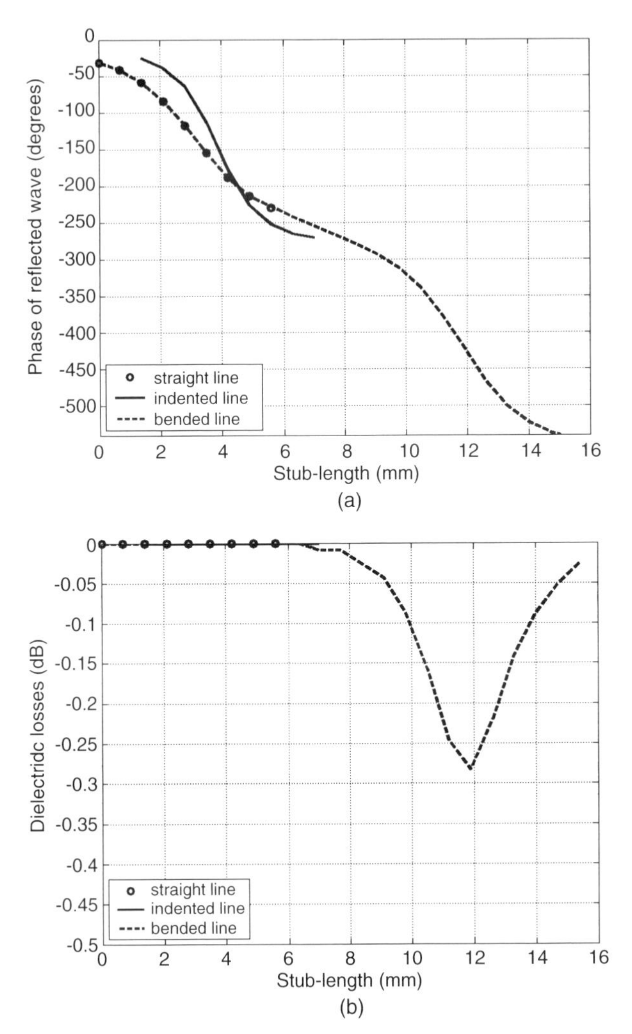

Figure 3.6. Reflected field for normal incidence versus the stub-length (t = 1.59 mm, $\varepsilon_r = 3.2$). (a) Phase, (b) dissipative losses.

3.5 FULL-WAVE ANALYSIS OF MULTILAYER PERIODIC STRUCTURES

For the analysis of a reflectarray assuming local periodicity, the phase-shift and losses can be computed independently for each element by analyzing only one cell in a periodic environment. As a result of the periodic boundary conditions in the cell, the field component can be expressed in any region of the periodic cell as a summation of Floquet modes, also called space harmonics. The discrete spectra of space harmonics simplifies the analysis problem when an integral equation is formulated and solved by Method of Moments, because the integrals are substituted by summations in the spectral domain [24–27]. As a result, the problem becomes similar to a problem of transversal discontinuities in waveguides, in which the field in each waveguide section is expressed as a summation of the eigen-modes.

The analysis technique for multilayer reflectarrays, such as those based on stacked patches or aperture-coupled stubs, can be addressed, either by an overall or a modular technique. In the overall method, the equations are formulated and solved for each specific periodic element as a whole, then the computation time and the number of unknowns increase rapidly with the number of array layers. On the other hand, a modular approach, as proposed in [26, 27, 43], is based on the computation of the Generalized Scattering Matrix (GSM) or any other multi-mode matrix for each array or interface layer. Then, the response of the whole multilayer cell is obtained through a simple cascading process, which only requires matrix operations.

The modular approach is very flexible for analyzing arbitrary geometries, since the matrix that characterizes each layer is computed independently and is used as a building block for the analysis of multilayer structures. Using this technique, the individual problems for each array or transition involves very simple two-layer Green's functions, and the number of unknowns does not increase with the number of layers. An additional layer only requires the solving of an additional two-layer problem plus simple matrix operations for the cascade process. Moreover, the framework of cascade connections allows us to take advantages of using different methods in computing the GSM for each surface of a multilayered structure. The modular technique described here has been successfully applied to the analysis of two different types of reflectarray elements; variable-sized stacked patches and aperture-coupled stubs.

The scattering parameters, that is, the elements of the GSM, can be computed either independently for each incident space harmonic [26], or at the same time for all space harmonics. In the first case, only a small number of space harmonics can be handled for the cascade process. This is not a limitation when the separation between two adjacent periodic surfaces is large, but for a small separation, hundreds of harmonics may be required in the cascade. For a periodic surface on a thin dielectric sheet, the problem can be solved by computing the GSM of the whole block, including the effect of the dielectric support in Green's function [26], but the analysis of a multilayer structure involving two close periodic surfaces through the cascade approach requires large GSMs.

In the technique described in this section, the GSM of a periodic surface on the interface between two dielectric media is computed and used as a building block. The incident field is assumed to be a summation of Floquet harmonics, instead of a single plane wave, and all the elements of the GSM are computed at once. The computer effort to obtain the complete GSM is almost equal to that required for a single scattering parameter, and the number of space harmonics for the GSM can be very large (several hundreds). This technique is very efficient for the analysis of multilayer structures with arbitrary geometries, especially when the distance between adjacent grids is small and a large number of space harmonics must be used in the cascade process, as is the case of multilayer reflectarrays.

3.5.1 Characterization of a Periodic Interface as a Building Block

A periodic array of conductive patches on the interface between two dielectric media, characterized respectively by the relative dielectric constants ε_{r1} and ε_{r2}, arranged in a rectangular lattice of dimensions $p_x \times p_y$, as shown in Fig. 3.7, is considered as the basic building block for the analysis of multilayer periodic structures (Fig. 3.8). In the reflectarray problem, the incident field coming from

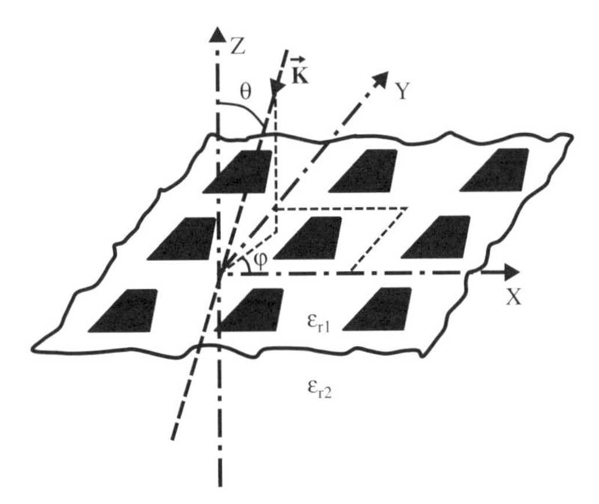

Figure 3.7. Periodic surface on the interface between two dielectric media used as building block.

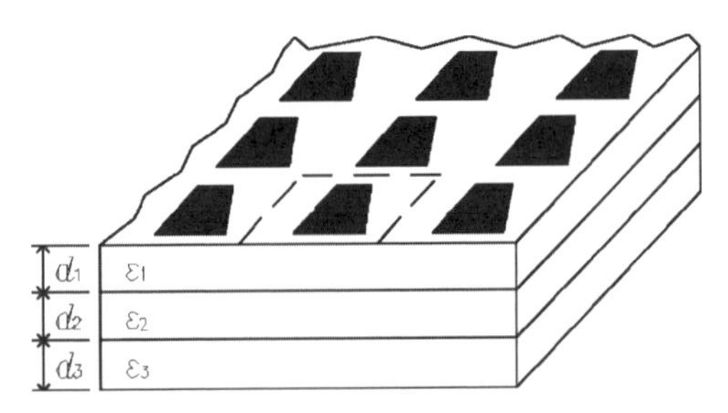

Figure 3.8. Multilayer periodic structure.

the feed on the first interface is always propagating in vacuum ($\varepsilon_{r1} = 1$), however, in the rest of the interfaces ε_{r1} and ε_{r2} should correspond to the dielectric layers. Without a loss of generality, the interface is located at $z = 0$ and the formulation in this section refers to this plane. The metallizations can be resistive and the dielectrics with losses. These effects are represented by a surface impedance Z_s for the metallizations and by complex relative permittivities ε_{r1} and ε_{r2} for the dielectrics.

An arbitrary incident field, expressed as a summation of Floquet space harmonics with complex amplitudes d_l, is considered on both sides of the periodic surface, instead of a usual single plane wave. From the upper side (medium characterized by ε_{r1}), the incident field is expressed as a summation of TE ($1 \le l \le L$) and TM ($L + 1 \le l \le 2L$) Floquet harmonics with amplitudes d_l. As is customary, subindex l includes all of the ordered pairs (m, n) of Floquet harmonics, ordered as the cut-off frequency increases. The transverse (to z) incident electric field propagating towards $-z$ is written as

$$\mathbf{E}_1^i = \sum_{l=1}^{2L} d_l \mathbf{e}_l \exp(j(k_{xm}x + k_{yn}y + k_{zl}z)). \tag{3.4}$$

Similarly, the transverse electric field in medium (1) propagating towards z, is expressed as

$$\mathbf{E}_1^r = \sum_{l=1}^{2L} a_l \mathbf{e}_l \exp(j(k_{xm}x + k_{yn}y - k_{zl}z)), \tag{3.5}$$

where a_l are the complex amplitudes (unknowns). This field includes the field reflected by the dielectric interface plus the field scattered by the metallic patches. The expressions of the normalized modal fields for the TE and TM Floquet harmonics $\mathbf{e}_l$ are given, respectively, as

$$\begin{cases} \mathbf{e}_l = \dfrac{1}{k_{cl}}(-k_{yn}\hat{\mathbf{x}} + k_{xm}\hat{\mathbf{y}}) & \text{for} \quad 1 \le l \le L \quad (TE) \\ \mathbf{e}_l = \dfrac{1}{k_{cl}}(k_{xm}\hat{\mathbf{x}} + k_{yn}\hat{\mathbf{y}}) & \text{for} \quad L+1 \le l \le 2L \quad (TM) \end{cases}$$

$$\text{with} \quad \begin{cases} k_{cl} = \sqrt{k_{xm}^2 + k_{yn}^2} \\ k_{xm} = k_0 \sin\theta\cos\varphi + 2m\pi/A = k_{x0} + 2m\pi/p_x \\ k_{yn} = k_0 \sin\theta\sin\varphi + 2n\pi/B = k_{y0} + 2n\pi/p_y. \end{cases} \tag{3.6}$$

k_0 is the free-space wavenumber, (θ, φ) the usual spherical angles characterizing the direction of the incident or radiated plane wave, that corresponds to the $l = 1$ ($m = n = 0$) space harmonic. For normal incidence, $\theta = \varphi = 0$, the inconsistency of Eq. (3.6) is solved by setting $k_{x0} = 1$ and $k_{y0} = 0$, which means that the TE wave is polarized with the electric field in the y-direction.

Note that k_{xm} and k_{yn} are defined for free-space, which is the first medium seen from the feed. They are unchanged in the rest of the layers because these components of the wave vector must be continuous. However, the z-component of the propagation vector varies with the dielectric constant, according to the expression

$$k_{zl1(2)} = \sqrt{\varepsilon_{r1(2)} k_0^2 - k_{xm}^2 - k_{yn}^2}, \tag{3.7}$$

where subindex 1(2) represents either medium (1) or (2).

The incident and scattered fields on the surface ($z = 0$), but seen from medium (2) characterized by ε_{r2} are expressed by summations of space harmonics similar to those in Eqs. (3.4) and (3.5), but with complex amplitudes c_l for the incident field (propagating in z direction), b_l for the scattered field propagating towards $-z$ direction.

The periodic surface **n** is characterized by its GSM, which includes all the higher order Floquet modes generated by the periodic array as

$$\begin{bmatrix} \mathbf{A}^n \\ \mathbf{B}^n \end{bmatrix} = \begin{bmatrix} \mathbf{S}_{11}^n & \mathbf{S}_{12}^n \\ \mathbf{S}_{21}^n & \mathbf{S}_{22}^n \end{bmatrix} \begin{bmatrix} \mathbf{D}^n \\ \mathbf{C}^n \end{bmatrix}, \tag{3.8}$$

where $\mathbf{D}^n$ and $\mathbf{C}^n$ are vectors with the amplitudes of the Floquet harmonics incident on the array ***n*** from the media (1) and (2), respectively. $\mathbf{A}^n$ and $\mathbf{B}^n$ are vectors with the amplitudes of the propagating and evanescent Floquet harmonics on the media (1) and (2), respectively, generated by the periodic array ***n*** and by the dielectric interface.

The GSM of each array of metallic patches is computed separately for each periodic interface by the Method of Moments in the spectral domain, but assuming a summation of Floquet modes as the incident field on both sides of the array. An incident field is considered simultaneously on both sides of the periodic surface, characterized by the vector $\mathbf{D}^n$ in medium (1) and by $\mathbf{C}^n$ in medium (2). However, by applying superposition, the problem can be formulated by considering the excitation from each side separately, and then the complete matrix is obtained directly by superposition. Assuming an incident field from medium (1), the equation to be solved by the Method of Moments is

$$\mathbf{E}_1^e(x, y) + \mathbf{E}_1^s(x, y) = Z_S \mathbf{J}_1(x, y), \tag{3.9}$$

where $\mathbf{E}_1^e(x, y)$ and $\mathbf{E}_1^s(x, y)$ are the excitation and scattered fields on the periodic surface, and $\mathbf{J}_1(x, y)$ is the current induced on the metallic patches. The excitation field is the total field produced on the dielectric interface without metallizations. It includes not only the incident field, but also the field reflected by the change of dielectric, as given by

$$\mathbf{E}_1^e(x, y) = \sum_{l=1}^{2L} d_l (1 + \Gamma_l) \mathbf{e}_l \exp(j(k_{xm} x + k_{yn} y + k_{zl} z)), \tag{3.10}$$

where Γ_l is the reflection coefficient at the dielectric interface for TE $(1 \le l \le L)$ and TM $(L + 1 \le l \le 2L)$ Floquet harmonics.

The scattered field is expressed as in [26],

$$\mathbf{E}_1^s(x, y) = \sum_{l=1}^{2L} \tilde{\mathbf{G}}(k_{xm}, k_{yn}) \bullet \tilde{\mathbf{J}}_1(k_{xm}, k_{yn}), \tag{3.11}$$

where $\tilde{\mathbf{G}}(k_{xm}, k_{yn})$ and $\tilde{\mathbf{J}}_1(k_{xm}, k_{yn})$ are, respectively, the associated dyadic Green's function and induced current over the metallizations, both in spectral domain.

When the metallizations are of rectangular or any other canonical shape, the current distribution can be expanded as a summation of entire-domain basis functions,

$$\mathbf{J}(x, y) = \sum_{\mathrm{p}=1}^{\mathrm{P}} \alpha_{\mathrm{p}} \boldsymbol{\psi}_{\mathrm{p}}^{\mathrm{h}}(x, y) + \sum_{\mathrm{q}=1}^{\mathrm{Q}} \alpha_{\mathrm{p+q}} \boldsymbol{\psi}_{\mathrm{q}}^{\mathrm{e}}(x, y), \tag{3.12}$$

in which $\boldsymbol{\Psi}_{\mathrm{p}}^{\mathrm{h}}$ and $\boldsymbol{\Psi}_{\mathrm{q}}^{\mathrm{e}}$ are, respectively, pth TE-mode and qth TM-mode basis functions. For noncanonical shapes, the metallic surface is divided up into a rectangular grid, and the $\boldsymbol{x}$ and $\boldsymbol{y}$ components of the current are expressed as a sum of subdomain basis functions similar to Eq. (3.12).

After substituting all the expressions in (3.9), and using a Galerkin testing procedure, the vector Λ containing the unknowns α_i $(i = 1, \mathrm{P} + \mathrm{Q})$ for the current distribution is obtained as a function of the excitation field (3.10), represented by the vector $(\mathbf{I} + \mathbf{R})\,\mathbf{D}$, where $\mathbf{I}$ is the identity matrix and $\mathbf{R}$ is a diagonal matrix with the reflection coefficients Γ_l. From the previous relation of vectors Λ and $\mathbf{D}$, the scattered field (3.11) produced by the induced currents is computed as a function of $\mathbf{D}$. After adding to the scattered field the field reflected by the dielectric interface, and identifying with Eq. (3.8) when $\mathbf{C}$ is equal to zero, the submatrix $\mathbf{S}_{11}^n$ is obtained using the procedure descried in [27] as

$$\mathbf{S}_{11}^n = \mathbf{S}_0^n(\mathbf{I} + \mathbf{R}) + \mathbf{R}. \tag{3.13}$$

The submatrix $\mathbf{S}_0$ is defined in [27], and it is obtained from the inverse of the impedance matrix in the Method of Moments and from other matrices involving the Green's functions, the basis functions in spectral domain and the modal functions (3.6). The element $\mathbf{S}_{11}(l_1, l_2)$ represents the complex amplitude of the Floquet mode l_1 reflected by the periodic surface when the mode l_2 is incident with unitary amplitude, and takes into account not only the fields produced by the induced currents, but also the fields reflected by the dielectric interface when the patches are not present. On the other hand, the element $\mathbf{S}_0(l_1, l_2)$ represents the complex amplitude of the Floquet mode l_1 in a medium (1) produced only by the induced currents on the patches for an excitation field (3.10) of unit amplitude corresponding to the incident Floquet mode l_2.

The boundary condition for the transverse electric field in the periodic cell results in

$$\mathbf{S}_{21}^{n} = \mathbf{I} + \mathbf{S}_{11}^{n} = (\mathbf{S}_{0}^{n} + \mathbf{I})(\mathbf{I} + \mathbf{R}). \tag{3.14}$$

Submatrices $\mathbf{S}_{22}^{n}$ and $\mathbf{S}_{12}^{n}$ of the $\boldsymbol{n}$ periodic interface are obtained by assuming the incidence from the medium (2). Because of the invariance of Green's function in the spectral-domain and the antisymmetry of reflection coefficients for the dielectric interface, these submatrices are obtained without additional computations, as described in [27],

$$\mathbf{S}_{22}^{n} = \mathbf{S}_{0}^{n}(\mathbf{I} - \mathbf{R}) - \mathbf{R}. \tag{3.15}$$

$$\mathbf{S}_{12}^{n} = \mathbf{I} + \mathbf{S}_{22}^{n} = (\mathbf{S}_{0}^{n} + \mathbf{I})(\mathbf{I} - \mathbf{R}). \tag{3.16}$$

The interfaces between two dielectric layers, without metallizations, can also be characterized by a GSM, which is obtained directly by analytically solving the problem of a plane wave under oblique incidence. However, when very thin dielectric layers are sandwiched in one or both sides of a periodic array, a high iteration of evanescent modes occurs, and the GSM must contain many space harmonics (several hundreds) to accurately account for that iteration. In this case, it is more efficient to include the effect of the thin dielectric layers in the Green's function, and to compute the GSM for the array of patches together with the thin substrate and/or superstrate, which will be considered as a building block for the whole multilayer structure [26].

Periodic surfaces with apertures in a ground plane can also be considered as a building block of the multilayer structure and represented by its GSM [30, 45]. In this case, the fields in the apertures can be expressed as a summation of the eigen modes of the waveguide with the same cross-section as the aperture, and then the Mode Matching technique is applied to compute the GSM. This technique has been applied in [45] for rectangular apertures and in [30] for arbitrarily shaped apertures. In the last case, the Finite Element Method was used to numerically compute the eigen modes associated with the noncanonical shape of the apertures.

3.5.2 Analysis of Reflectarray Elements in a Periodic Environment

For a reflectarray, the multilayer structure is normally backed by a ground plane to ensure the total reflection of the energy incident from the feed. If the ground plane is considered as a perfect conductor, its GSM is directly minus the identity matrix, because all incident space harmonics are reflected with a reflection coefficient equal to –1. The ohmic losses in the ground plane can be easily included in the scattering matrix by considering the real amplitude of the reflection coefficients. This evaluation can be used if the ground plane is implemented by a carbon–fiber–reinforced plastic (CFRP) layer, normally used in space applications, which typically adds 0.1 dB of ohmic losses.

The total GSM for a multilayered structure, as shown in Fig. 3.8, is obtained by the iterative cascading of the individual GSMs of each interface using simple matrix computations, by a process described in [27]. The analysis method is summarized in the diagram shown in Fig. 3.9. To obtain the GSM of a multilayer structure, the cascade connection begins with the GSMs of the first two surfaces, separated by a dielectric layer of thickness t_1. Then, the resulting matrix GSM for the sandwich with two interfaces is used to cascade with the GSM of the next surface and so on, as is illustrated in Fig. 3.10.

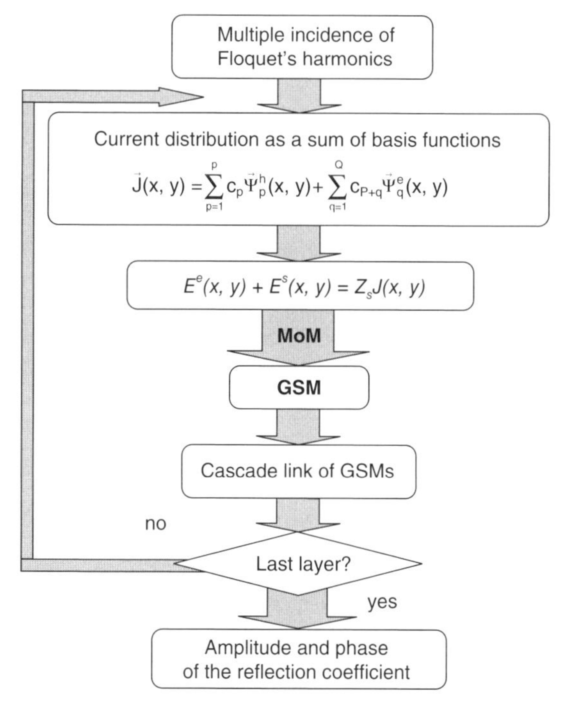

Figure 3.9. Diagram of the analysis technique proposed for studying multilayer periodic structures.

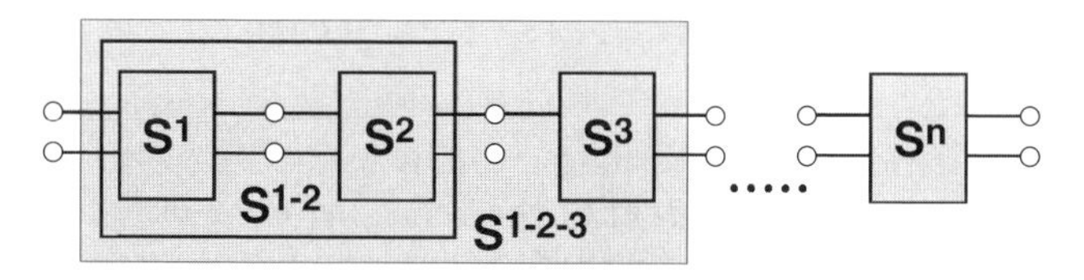

Figure 3.10. Linking process of the GSMs (cascade process).

The overall matrix $\mathbf{S}^{\mathrm{T}}$ is computed for N_s TE plus N_s TM Floquet modes, and fully characterizes the electromagnetic response of the multilayer periodic cell, including the phase-shifts for both polarizations, dissipative losses, and generation of cross-polarization. If there is no ground plane, or when there are apertures in the ground plane, as can be in the aperture-coupled element, the submatrix $\mathbf{S}_{21}^{\mathrm{T}}$ gives the power transmitted to the opposite layer. However, in the most common configuration, a ground plane is used and no transmission is produced, then the GSM (3.8) is reduced to

$$\mathbf{A} = \mathbf{S}_{11}^{\mathrm{T}}\mathbf{D}. \tag{3.17}$$

The reflection coefficient for TE and TM polarized waves are, respectively, the elements (1,1) and (L + 1, L + 1) of the submatrix $\mathbf{S}_{11}^{\mathrm{T}}$. The coefficients that generate cross-polarization, that is, reflection of TE when TM is incident, or vice versa, are the elements (1, L + 1) or (L + 1, 1) of $\mathbf{S}_{11}^{\mathrm{T}}$, respectively. An arbitrary field incident on a periodic cell will be broken down into TE and TM components, and will be defined by the amplitudes d_1 and d_{L+1} according to Eqs. (3.4) and (3.6), with the rest of the complex amplitudes of vector **D** equal to zero. The complex amplitudes a_l of the reflected TE and TM Floquet harmonics are computed by substituting **D** in (3.17). The reflected tangential electric field at a reflectarray cell (z = 0) for the multilayer periodic cell is obtained by substituting the coefficients a_l in (3.5). The tangential magnetic field is also obtained from coefficients a_l as

$$\mathbf{H}_1^r = \sum_{l=1}^{2L} a_l \mathbf{h}_l \exp(j(k_{xm}x + k_{yn}y - k_{zl}z)), \tag{3.18}$$

where $\mathbf{h}_l$ are the normalized tangential magnetic fields for the TE and TM Floquet harmonics [26], obtained from $\mathbf{e}_l$ by

$$\mathbf{h}_l = \frac{1}{\eta_l}\mathbf{z} \times \mathbf{e}_l, \tag{3.19}$$

where η_l is the TE or TM modal impedance.

$$\begin{aligned} \eta_l &= \frac{\omega\mu}{k_{zl}} \quad \text{for} \quad 1 \le l \le L \quad (TE); \\ \eta_l &= \frac{k_{zl}}{\omega\varepsilon} \quad \text{for} \quad L+1 \le l \le 2L \quad (TM) \end{aligned} \tag{3.20}$$

The tangential electric or magnetic field given by Eq. (3.5), (3.18) on each cell can be used to compute the far field of the reflectarray by integration of the field on the elementary cells. However, if the period has been chosen to avoid the propagation of higher-order space harmonics, as is customary, only the fundamental TE and TM modes associated with the incident plane waves, $l = 1$ and $l = \mathrm{L} + 1$, respectively, contribute significantly to the radiation patterns. Therefore, the complex coefficients a_1 and a_{L+1} can be used for the computation of the radiation patterns, as it will be discussed later. Note that the coefficients for the reflected field includes the coupling between TE and TM polarizations from the incident to the reflected field, as follows:

$$a_1 = S_{11}^T(1,1)d_1 + S_{11}^T(1,\mathrm{L}+1)d_{L+1} \tag{3.21}$$

$$a_{\mathrm{L}+1} = S_{11}^T(\mathrm{L}+1,1)d_1 + S_{11}^T(\mathrm{L}+1,\mathrm{L}+1)d_{L+1} \tag{3.22}$$

The method of analysis uses the usual TE and TM decomposition, as defined in Eq. (3.6). However, when a reflectarray is designed for dual-linear or circular polarization, two orthogonal electric field components are considered on the aperture of the feed horn, one in $\boldsymbol{y}$ direction ($\boldsymbol{y}$-polarization) and the other with $\boldsymbol{x}$ and $\boldsymbol{z}$ components of the electric field ($\boldsymbol{x}$-polarization), assuming the coordinate system detailed in Fig. 3.1.

To control the phase of these two independent components of the field coming from the feed, two geometrical parameters must be adjusted independently on the reflectarray, in orthogonal directions, such as the length of stubs or the patch dimensions. To compute the phase-shift produced for each polarization of the feed, and also the cross-polarization, the $\boldsymbol{x}$- *or* $\boldsymbol{y}$-polarized field incident on each reflectarray cell must be expressed as a linear combination of the classic TE and TM fields. The complex amplitudes of the incident TE and TM plane waves that correspond to the $l = 1$, L + 1 ($m = n = 0$) space harmonics in Eq. (3.4), are related to the $\boldsymbol{x}$ and $\boldsymbol{y}$ components by

$$\begin{pmatrix} d_1 \\ d_{L+1} \end{pmatrix} = \frac{1}{k_{cl}} \begin{pmatrix} -k_{y0} & k_{x0} \\ k_{x0} & k_{y0} \end{pmatrix} \begin{pmatrix} d_x \\ d_y \end{pmatrix}. \tag{3.23}$$

The coefficients a_1 and a_{L+1} of the reflected field are obtained by (3.21 and 3.22), and are transformed again into $\boldsymbol{x}$ *and* $\boldsymbol{y}$ components on the surface of the reflectarray

$$\begin{pmatrix} a_x \\ a_y \end{pmatrix} = \frac{1}{k_{c1}} \begin{pmatrix} -k_{y0} & k_{x0} \\ k_{x0} & k_{y0} \end{pmatrix} \begin{pmatrix} a_1 \\ a_{L+1} \end{pmatrix}. \tag{3.24}$$

The transformations relating TE-TM with $\boldsymbol{x}$-$\boldsymbol{y}$ decomposition given by (3.23 and 3.24) apply not only to the fundamental Floquet modes, but also to higher-order modes, if k_{x0} and k_{y0} are substituted by k_{xm} and k_{yn}, respectively. These transformations can be introduced into Eq. (3.17) in order to obtain an $\mathbf{S}^{\mathrm{C}}$ matrix relating the Cartesian components of incident and reflected fields in the periodic cell

$$\mathbf{A}^C = \mathbf{S}_{11}^C \mathbf{D}^C. \tag{3.25}$$

For the characterization of a reflectarray element, the phase-shift and dissipative losses must be computed for each polarization as a function of the geometrical parameter used to control the phase-shift. The phase-shift and losses are directly the phase and amplitude of the elements (1, 1) and (L + 1, L + 1) of the submatrix $\mathbf{S}_{11}^T$ for TE and TM components, or $\mathbf{S}_{11}^C$ for Cartesian components. Note that for normal incidence ($k_{x0} = 1$, $k_{y0} = 0$) the TE wave corresponds to a $\boldsymbol{y}$-directed electric field, and the TM to an $\boldsymbol{x}$-directed field, but field transformations (3.23 and 3.24) are required for oblique incidence. The dissipative losses will include the losses in all the dielectric layers modeled by a complex permittivity, and also the losses in the conductors modeled by a surface impedance Z_S as shown in (3.9).

The previously described modular technique has been applied to the analysis of two types of reflectarray elements; the first one based on variable-sized patches, and the second one based on aperture-coupled stubs. The results are discussed in the following sections.

3.6 PHASE-SHIFTER ELEMENT BASED ON SINGLE AND STACKED VARIABLE-SIZED PATCHES

The phase adjustment by the variation of the resonant length of printed patches, as shown in Fig. 3.11 for square patches, is very easy to implement by using printed circuit technology. This phasing technique eliminates some of the inconveniences associated with the stubs. First, the problem of accommodation of the stubs with lengths of up to half-a-wavelength is eliminated. Second, the deterioration of some electrical performances, such as dissipative losses and cross-polarization produced by the bent stubs, are reduced in this implementation.

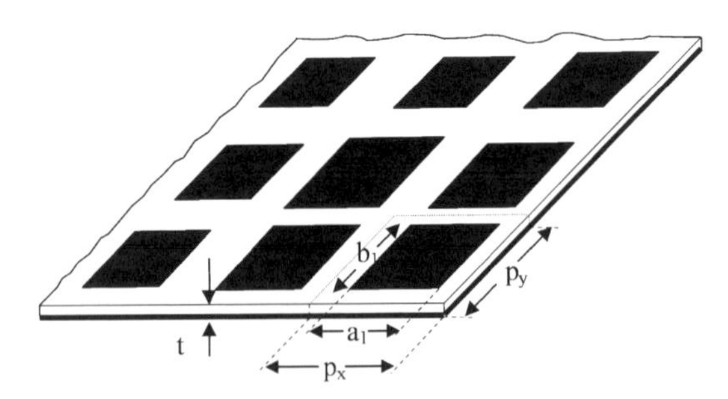

Figure 3.11. Rectangular patches of varying size on a grounded dielectric slab.

3.6.1 Single-Layer Elements

The operating principle of the reflectarrays of variable-sized printed elements is based on the fact that the phase of the reflected wave varies with the resonant length of the elements. A microstrip patch is a resonant antenna, so that its length should be approximately half a wavelength in the dielectric. If the patch length is modified in an array of rectangular patches on a grounded dielectric, as shown in Fig. 3.11, the phase of the reflected field will be changed. For a resonating patch, a small change in its size produces a wide range in phase variation of the reflected wave, primarily due to the high Q nature of the patch antennas [6]. The amplitude of reflection coefficient must be nearly equal to one, provided that there is no grating lobe or surface wave generation, because of the ground plane. A small reduction in amplitude is produced by the dissipative losses in the dielectric separators and on the metal patches.

The mechanism for controlling the phase of the reflected wave by varying the resonant dimensions of microstrip patch antennas was reported for first time in [5] for crossed dipoles and in [6] for rectangular patches. Empirical phase curves obtained by measuring the phase-shift versus the length of the cross dipoles where shown in [5]. However, for the analysis and design of a reflectarray, it is more practical to obtain the phase-shift through electromagnetic simulations. A full-wave technique based on MoM was proposed in [6] to obtain the phase curves as a function of the patch dimensions, by considering the incidence of a plane wave on an infinite array of rectangular patches. This approach takes into account the mutual coupling effects between the reflectarray elements, as well as the specular reflection from the ground plane supporting the patches.

The total range of phase variation that can be achieved by varying the length of the patches depends on the separation between patches and ground plane, that is, the thickness of the substrate. For thickness smaller than a tenth of wavelength, a 330° range can be achieved, which is enough for carrying out practical designs, but this range diminishes for thicker substrates. Because of this, the reflectarrays based on this adjustment technique use thin dielectric substrates. However, the phase variation versus the length is strongly nonlinear because of the narrow band behavior of microstrip patches, exhibiting very rapid variations near the resonance, and very slow in the extreme values, as can be seen in Fig. 3.12. The rapid phase variation makes the phase distribution very sensitive to manufacturing tolerance errors. Because of the nonlinear behavior, the phase is very sensitive to variations in frequency, significantly reducing the working band of the reflectarray.

A way to achieve a smoother behavior of the phase as a function of the length consists of increasing the thickness of the substrate, but this significantly reduces the total phase range. It must be kept in mind that for the design of a reflectarray, all values of phase-shift in a range of 360° are required, and they cannot be achieved for a thicker substrate. One possible solution to achieve a

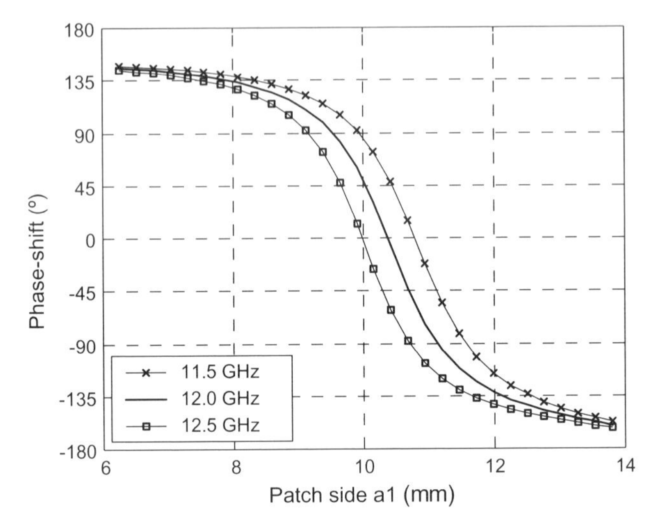

Figure 3.12. Phase-shift at normal incidence for a periodic array of square patches on a grounded substrate versus the patch side a_1 at three frequencies ($p_x = p_y = 14$ mm, $t = 1$ mm, $\varepsilon_r = 1.05$).

more linear behavior of the phase versus size in a range wider than 360°, is to stack two or more array layers, as is explained below.

3.6.2 Multilayer Elements

It is well known that an array of rectangular metallic patches behaves as a resonant circuit, in which the phase of the reflected field varies with the size of the patches within a range of up to 180°. When the array is backed by a metallic plane, as in Fig. 3.11, the maximum phase shift range approaches 360°, if the separation between the patches and the plane is very small compared with the wavelength, see Fig. 3.12. However, when two or more array layers are used as in Fig. 3.13, each of them behaves like a resonant circuit, and the phase of the reflected field varies with the patch size in a similar way to that of one layer, but the phase shift can reach values of several times 360°. Therefore, with several array layers, the separation between them, and the separation between the last array and the metallic plane, can be increased to achieve a smoother and more linear behavior of the phase as a function of the patch size, maintaining a range for phase shift greater than 360° (Fig. 3.14).

The multilayer structure with rectangular patches shown in Fig. 3.13 has been used as a phase-shifter element as reported in [14–16]. The phase control is carried out by modifying the dimensions of the metallic patches. However, the phase of the reflection coefficient will be dependent not only on the size of the elements but also on the angle of incidence (θ_i, ϕ_i) and type (E or H)

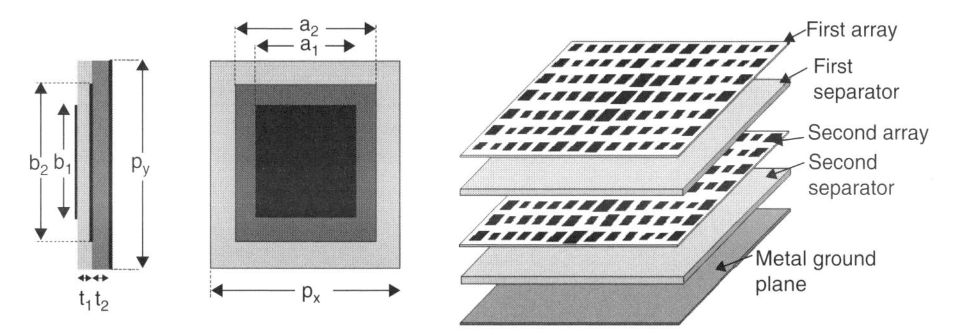

Figure 3.13. Two-layer reflectarray. (a) Periodic cell, (b) multilayer structure.

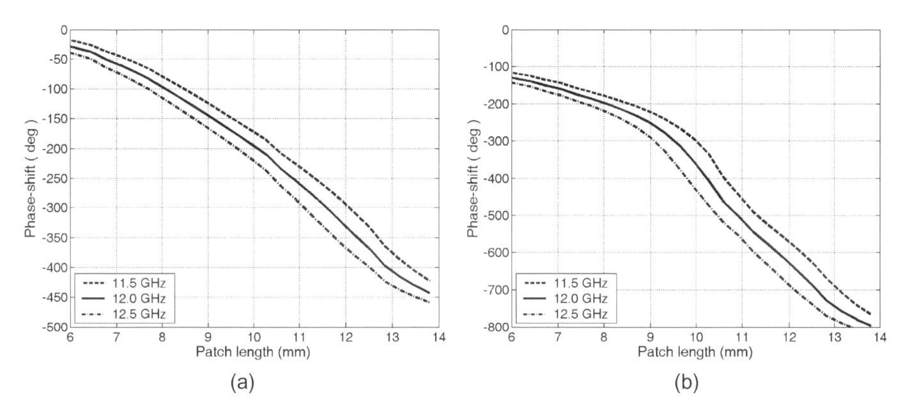

Figure 3.14. Phase-shift at normal incidence for a multilayer periodic structure, defined in Fig. 3.13(a), versus the patch side of the array closer to the ground plane. ($a_1 = b_1$, $a_2 = b_2$, $p_x = p_y = 14$ mm, $t_1 = t_2 = 3$ mm, $\varepsilon_r = 1.05$). (a) Two layers ($a_1 = 0.7a_2$), (b) three layers ($a_1 = 0.7a_3$, $a_2 = 0.9a_3$).

of polarization. This phase-shifter element can be used for dual linear or circular polarization. To consider any kind of polarization, the field at each element is broken down into two components parallel to the patch sides, and the phase of each polarization is controlled by adjusting the corresponding dimensions of the stacked rectangular patches. The multilayered periodic structure is analyzed using the previously described modular technique.

As an example, the phase of the reflection coefficient is computed for both orthogonal polarizations, assuming square patches in a periodic cell of 14 × 14 mm, and assuming a fixed ratio of patch sizes on each layer. The dielectric between each array layer is Rohacell®, 3-mm thick. Fig. 3.14 shows the phase of the reflection coefficient as a function of the patch side, at three different frequencies, for periodic structures of two and three layers. In this case, a normal incidence has been considered and the phase in both polarizations is identical. For the two-layer structure, the phase variation is very linear in a 360° range and the behavior is similar at the three frequencies. This means a significant improvement in the bandwidth of the element. In addition, the phase slope vs. dimensions (65°/mm) is much lower than in the phase response for a single layer (135°/mm). Then, a tolerance error in patch dimension of 0.1 mm will produce only an error in phase of 6.5°, which means a low sensitivity to manufacturing tolerances.

On the other hand, the three-layer structure produces larger values of phase slope and wider phase range (two times 360°). The three-layer phase-shifter allows more flexibility for optimizations at several frequencies [16], as will be discussed in a later chapter. For the three-layer element, the phase-shift is shown in Fig. 3.15 for different angles of incidence and polarization. The differences in phase-shift for the two orthogonal polarizations, with the electric field in the ***x***- and ***y***-direction, can be larger than 60° for an incidence of 40° ($\theta_i = 40°$, $\phi_i = 0°$), and the phase-shift also differs from that corresponding to normal incidence, particularly for ***x***-polarization. These results show that for a more accurate analysis, the real angle of incidence and polarization of the field must be taken into account.

3.6.3 Measurements of Phase-Shift and Losses in Waveguide Simulator

Phase-shift and dissipative losses of the reflectarray element are predicted by the electromagnetic simulations, because they are directly the phase and amplitude of the reflection coefficient, computed when a plane wave is inci-

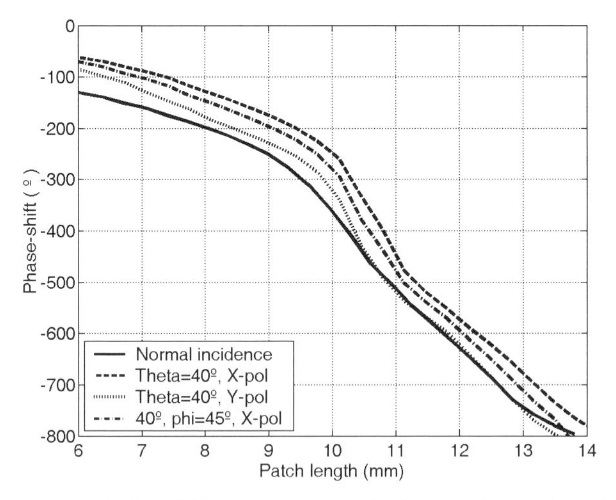

Figure 3.15. Phase-shift for different angles of incidence for a three-layer periodic structure, versus the patch length a_3.

dent. Phase-shift and losses can also be measured using small samples in a waveguide simulator (WGS). The WGS model is a well-known technique used to measure the active element performance of array elements in the array environment [46]. The WGS is based on the fact that if one or more cells with printed patches are inserted into a rectangular waveguide, the behavior of an infinite array is achieved for a given angle of incidence, as a result of the image theory. The fundamental mode TE_{10} of the rectangular waveguide is the superposition of two plane waves propagating on the H-plane in symmetric directions, $\pm\theta_0$, with respect the waveguide axis, so that the tangential electric field is cancelled on the lateral metallic walls. Then, the reflection coefficient measured for the TE_{10} mode corresponds to that of a plane wave incident on the infinite array at the angle θ_0, which is related to the frequency by the propagation constant of the TE_{10} mode. The main limitations in the measurements in WGS are: 1) they are only applicable to a single linear polarization, and 2) the angle of incidence is determined by the element spacing and the frequency, so the WGS only represents an infinite array for a particular combination of frequency and angle.

To check the accuracy of the analysis tool for the multilayer reflectarray element, several samples have been manufactured and tested in WGS. The samples consists of two elements of the three-layer reflectarray placed in a WR90 waveguide (dimensions 22.86 × 10.16 mm), for measurements in the X-band. The reflection coefficient is measured using an HP 8510B network analyzer and also computed using the analysis routine, taking into account the angles of incidence corresponding to each frequency. Three samples have been manufactured: without patches, with small patches ($a_3 = a_2 = 7$, $a_1 = 6$ mm), and with large patches ($a_3 = a_2 = 8.4$, $a_1 = 7.5$ mm) (Fig. 3.16). The patches in each layer are printed on a 25-micron-thick Kapton film, which is bonded to a Kevlar fabric composite layer using low-loss Cyanate-Ester resin and sepa-

Figure 3.16. Samples for wavequide simulator measurements.

rated by 2-mm-thick Rohacell HF51. Simulated and measured values of losses and phase-shift are compared in Fig. 3.17.

For the samples without patches, the two curves of phase-shift on the top of Fig. 3.17(a) cannot be distinguished. For the other samples, phase-shift becomes more negative in Fig. 3.17 for larger patches, and a good agreement is observed between measurements and simulations. The dissipative losses are only represented for the case with large patches, in which the losses are larger because the patch dimensions are near resonance at higher frequencies. The

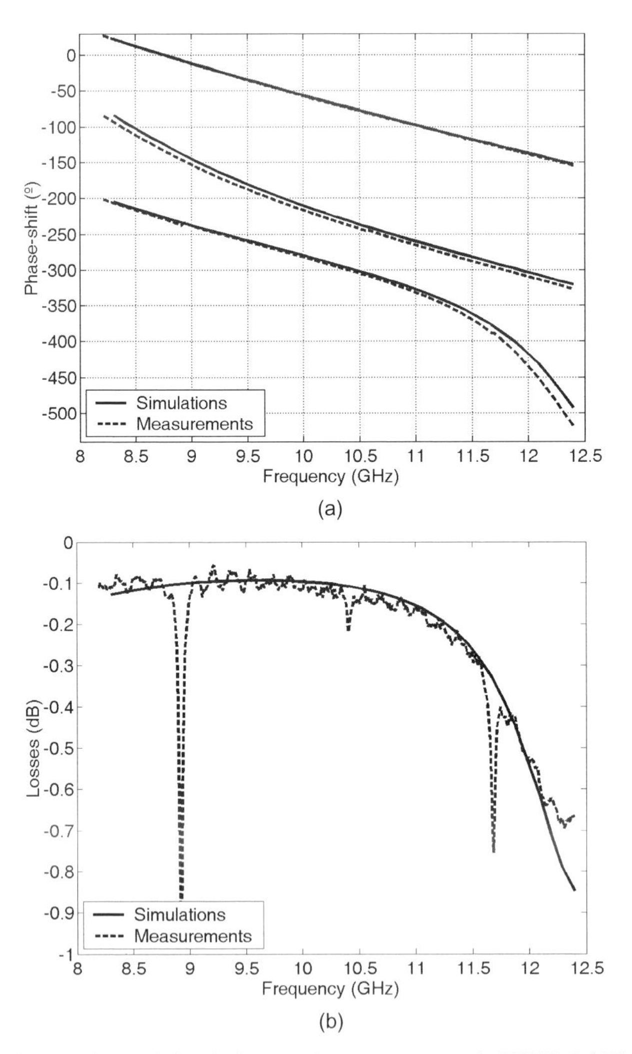

Figure 3.17. Comparison of simulations and measurements in WGS. (a) Phase-shift for three samples, (b) losses for large patches.

sharp augmentation of losses at 8.9 GHz and at 11.7 GHz is produced by resonances in the cavity of the WGS, and for this reason are not reproduced in the simulations.

3.7 PHASE-SHIFTER ELEMENT BASED ON APERTURE-COUPLED PATCHES

Aperture-coupled patches with stubs of different lengths have been used for single or dual linear polarization reflectarrays [9, 10]. In this configuration, each stub is made up of an open-ended length of microstrip line on the opposite side of the ground plane, which is electromagnetically coupled to the radiating patch by an aperture in the ground plane as show in in Fig. 3.18. The principle of operation is as follows: when the plane wave representing the field coming from the feed impinges on the resonant square patch, it is coupled to the microstrip line on the bottom (dipole) and propagates until reaching the open circuit extreme where it is reflected, coupled, and reradiated by the patch. In this configuration, as in that of the attached stubs, the phase of the reflected field should be proportional to twice the length of the stub, and a real phase delay can be achieved. The range of phase delay can be larger than 360° and is only limited by the maximum length of the stub. By printing the stubs in a different layer than that of the radiating elements, there is more room for the line, so the range in phase delay can be increased. The spurious radiation produced by the stubs is in the opposite direction with respect to the reflectarray main beam and can be eliminated by placing a ground plane underneath the lines, as shown in Fig. 3.18(a). An additional advantage of this configuration is

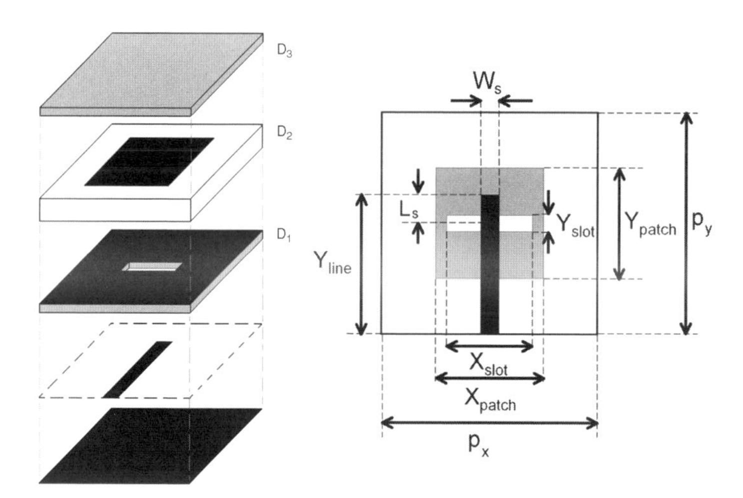

Figure 3.18. Aperture-coupled periodic cell. (a) Expanded view, (b) top view.

that active elements or controllable phase-shifters can be included in the microstrip line for reflectarrays with reconfigurable or steerable beams [47].

The aperture-coupled element must be designed in order to achieve a linear phase response in a range larger than 360°, as in the previous multilayer configuration. For the design of the phase-shifter, the microstrip line or dipole underneath the aperture (Fig. 3.19), is considered as two segments seen from the aperture center, the first one is a stub of a fixed length, which is adjusted to match the radiating element, whereas the second has variable length and is used to control the phase-shift.

First the aperture-coupled radiating element must be designed to achieve a good match between the line and the patch, considering a microstrip feed line as in the case of conventional planar arrays (Fig. 3.19(a)). At this step, by using a simulation tool, all the geometrical parameters, such as dielectric materials, thickness, dimensions of the aperture, patch, and fixed stub, have to be determined. The matching of the radiating element is important to allow the energy coming from the feed to be coupled to the delay line, according to reciprocity theorem. Once the aperture-coupled radiating element is designed, the curves of phase-shift are obtained by computing the phase of the reflection coefficient for different dipole lengths when a plane wave is incident on the patches (Fig. 3.19(b)).

3.7.1 Design of Reflectarray Element

Some simple equivalent circuits have been proposed for the aperture-coupled patch [48, 49], but they do not provide sufficient accuracy for the design of the radiating element. This is because the input impedance is very sensitive to several parameters, such as the aperture dimensions. For both, the design of the radiating element and for the computation of the phase-shift curves, it is

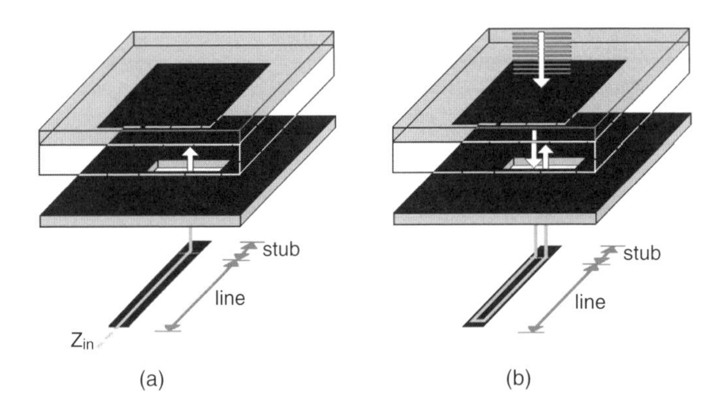

Figure 3.19. Aperture-coupled element. (a) Model for designing the radiating element, (b) model for computing the phase-shift curves.

preferable to use a full-wave simulation tool assuming the "infinite array model." Since the apertures and patches are identical, and the reflectarray usually comprises a very large number of elements, the reflectarray can be considered as a periodic array, with the exception of the stubs of different lengths. The technique based on the Method of Moments applied to the infinite array model, proposed in [25] and extended in [43] for multiple dielectric layers and stacked patches, is very accurate for the analysis of the radiating element in the array environment, because it takes into account all possible coupling effects between apertures and patches. The coupling between the stubs is ignored, but this is less significant.

The technique described in [43] is a modular approach that consists of characterizing each array or interface by a matrix (GSM). The GSM that characterizes the transition from the microstrip line to the aperture in an array environment is computed using the technique described in [43]. For this matrix representation, only the fundamental quasi-TEM mode in the microstrip is considered, and the GSM relates the incident and scattered mode in the mocristrip with all the Floquet harmonics on the layer above the array of apertures. The array of rectangular patches and the dielectric interfaces are characterized by their GSMs as explained in Section 3.5. Then, the individual matrices are cascaded in an iterative process as described in Section 3.5. The S_{11} element of the resulting matrix for the whole periodic cell shown in Fig. 3.19(a) gives directly the reflection coefficient at the microstrip port, which should be small enough for a good matching of the radiating element.

For the design of the radiating element, the process is similar as in conventional printed array antennas. First, the dielectric layers are defined and the dimensions are adjusted until a good impedance matching is achieved. Although interaction among all the parameters occurs, the basic ideas for the adjustment of the dimensions are as follows: the square patch is used to adjust the resonant frequency, the rectangular aperture is modified until the desired input resistance is achieved, and finally the fixed stub is used to compensate the imaginary part of the input impedance.

A radiating element has been designed to achieve return losses from the microstrip line of less than –22 dB in the frequency band 9–10 GHz [50], using the analysis method described in [43]. Return losses and the input impedance at the microstrip are shown in Fig. 3.20 for a radiating element defined in Table 3.1 in a period of 20 × 20 mm.

3.7.2 Phase Delay Curves

Once the radiating element has been designed, and the dimensions of the patch, aperture, and matching-stub have been determined, the structure is analyzed as a multilayer periodic structure to compute the phase-shift curves when a plane wave is incident, by using the modular approach described in Section 3.5. As in the case of stacked patches, the technique allows dissipative losses and phase-shift to be computed for any angle of incidence and for an

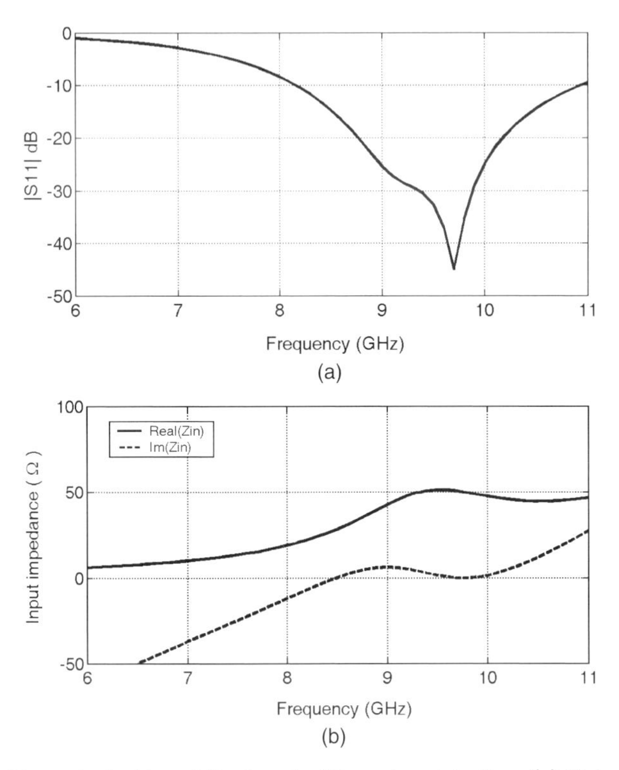

Figure 3.20. Characterization at the input of the microstrip line. (a) Return losses, (b) input impedance.

TABLE 3.1. Geometrical definition of reflectarray element

Layer	ε_r	$\tan\delta$	Thickness (mm)	X (mm)	Y (mm)
Line	—	—	0.00	$W_s = 1.85$	$L_s = 2.7$
D1	3.2	0.0030	0.79	—	—
Slot	—	—	0.00	6.0	1.0
D2	1.05	0.0002	2.00	—	—
Patch	—	—	0.00	10.0	10.0
D3	3.2	0.0030	0.79	—	—

arbitrary polarization of the incident field. The cross-polarization components are also computed, and are used to predict the cross-polarization patterns of the reflectarray.

For the reflectarray element defined in Table 3.1, the amplitude and phase of the reflection coefficient for normal incidence are shown in Fig. 3.21 as a function of the line length, for a periodic cell without a ground plane. Although

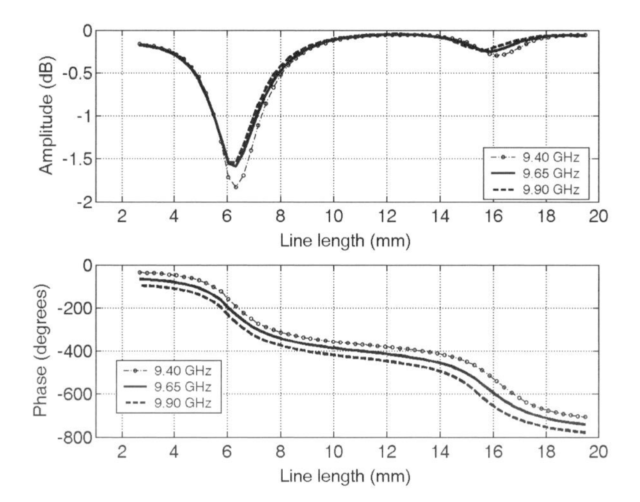

Figure 3.21. Amplitude and phase of the reflection coefficient for the periodic cell as a function of the dipole length for 50Ω line and $\varepsilon_{rs1} = 3.2$.

the phase delay should be a linear function of the stub length, the behavior is modified as a result of the resonances of the structure. At the two resonances observed in Fig. 3.21, for a line length in the order of 6.2 mm and 16 mm, the phase slope is changed and the amplitude of the reflection coefficient is slightly reduced, because a fraction of the energy is transmitted to the lower layer, particularly for dipoles with a length of around 6.2 mm. To eliminate the transmission near resonances, a ground plane has been placed at a distance $\lambda/4$ from the stub, as shown in Fig. 3.18(a). The amplitude and phase of the reflection coefficient obtained with and without a ground plane are compared in Fig. 3.22. The results show that the phase-shift curve is unchanged and the reflection losses are lower than 0.1 dB for all dipole lengths, which represent the dissipative losses in the dielectric layers. The phase curves are represented in Fig. 3.21 for three frequencies and the difference is practically uniform, therefore a good bandwidth of the phase-shifter is provided.

For the experimental validation of the phase-shifter element, the aperture-coupled radiating element previously designed in an array environment was redesigned for the waveguide simulator, considering two cells in a WR112 waveguide. In this case, the dielectric D3 was substituted by 0.508-mm thick GIL with $\varepsilon_r = 3.38$ and tan $\delta = 0.005$. The patch dimensions were unchanged, and the slot and stub were changed as follows: $X_{slot} = 6.65$ mm, $Y_{slot} = 1.2$ mm, $W_s = 1.6$ mm, $L_s = 2.06$ mm. The resulting input impedance from the microstrip was approximately equal to 50Ω in the frequency band 7.5–8.5 GHz. The patches and dielectric layers D2 and D3 were inserted in the waveguide, where the ground plane with the apertures are in contact with the waveguide flange, so that the delay lines were out of the waveguide (Fig. 3.23). The comparison

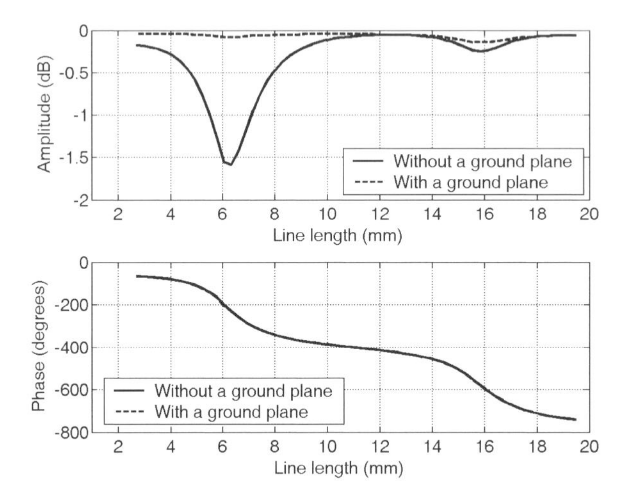

Figure 3.22. Amplitude and phase of the reflection coefficient for a reflectarray unit cell.

Figure 3.23. Samples for measurements in WR112 waveguide simulator.

between the calculated and measured phase of the reflection coefficient is shown in Fig. 3.24 for the several frequencies as a function of the line length. As shown in the figure, a good agreement has been obtained between the theoretical predictions and measurements in a range close to 360° of phase delay. The small differences between measurements and simulations are within an acceptable range and can be due to both manufacturing tolerances and the assumption of an infinite array of delay lines in the simulations. It must be

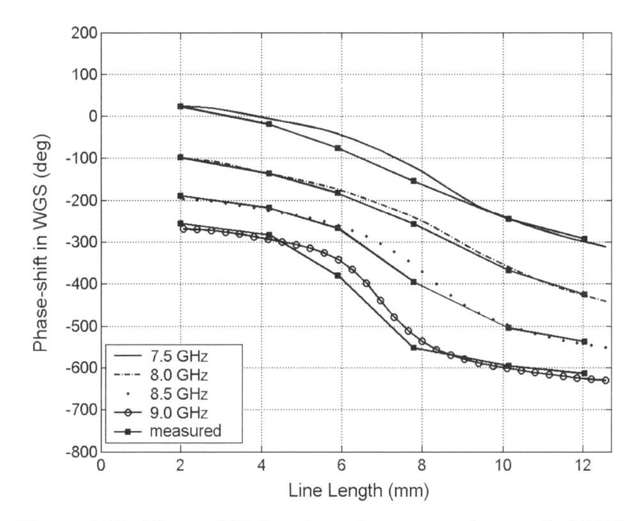

Figure 3.24. Phase-shift for the reflectarray element in WGS.

noted that the phase is practically linear at central frequency, and the behavior is smooth at the other frequencies.

The measured and simulated reflection losses are shown in Fig. 3.25 at the central frequency (8.00 GHz). The measured losses are in close agreement with the simulations, but they are very high (>1 dB). The reason for these high losses is that the energy is radiated by the apertures, to the opposite side, known as back-radiation in aperture-coupled patch arrays. This radiation can be eliminated as previously discussed by placing a ground plane separated $\lambda/4$ from the delay lines. The simulated losses with ground plane are also superimposed in Fig. 3.25 and the results show that only dissipative losses in the order of 0.3 dB are present. The presence of the ground plane practically does not affect the phase performance.

The total range in phase delay can be increased by increasing the electrical length of the delay line, which means increasing the physical length, the dielectric constant of the layer D2, or both [51]. For a larger increase in the line length, the line can be folded as shown in Fig. 3.26, to avoid the stub going out of the unit cell. In addition to increasing the range of phase delay, it is important to improve the linearity of phase delay versus dipole length. Although the phase delay should be directly proportional to the dipole length, a nonlinear behavior has been seen in previous results produced by the resonances of the dipole. The linearity of phase delay can be improved by modifying the length and width of the matching stub as shown in Fig. 3.26(b). Reflection losses and phase delay are shown in Fig. 3.27 at the central and extreme frequencies for the phase-shifter element defined in Table 3.1, but with the dimension of matching stub, L_s = 5 mm, W_s = 4 mm, and with a U-shaped line 1.85 mm wide, as in Fig. 3.26. The phase delay is greater than four 360° cycles. The amplitude

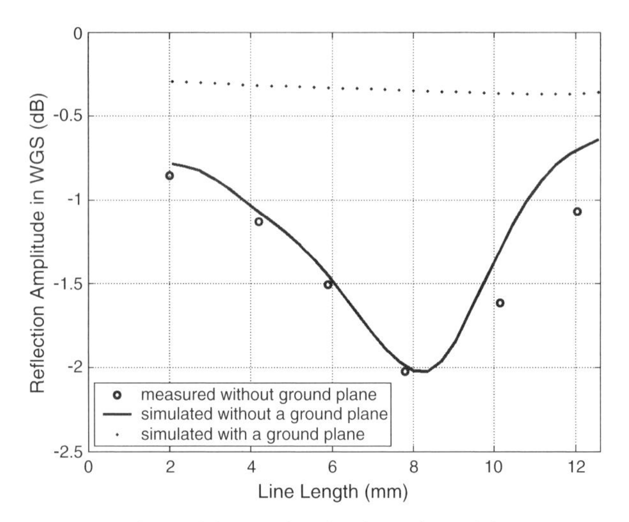

Figure 3.25. Reflection loses in WGS.

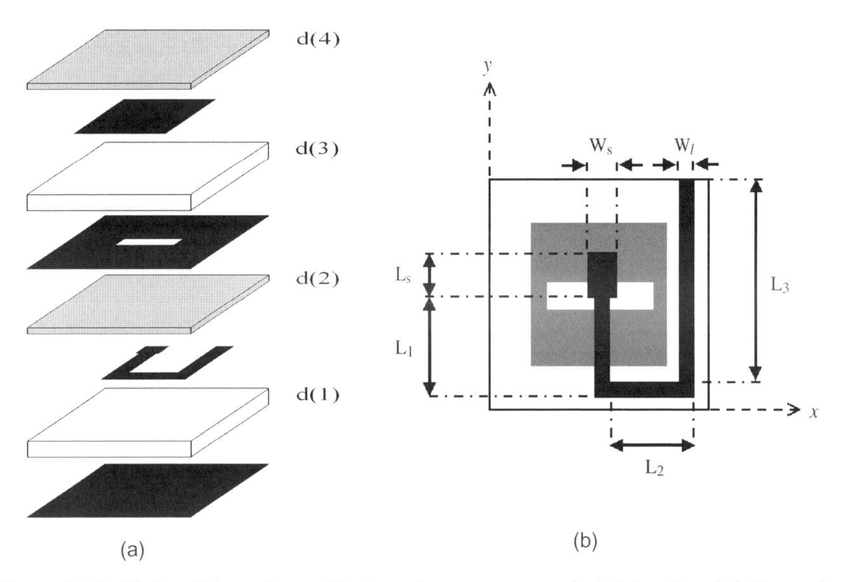

Figure 3.26. Unit cell based on a U-shaped aperture-coupled delay line (a) Expanded view, (b) top view. (*E. Letters* 2006, reprinted with permission.)

of reflection coefficient, assuming a ground plane at 7.7 mm, shows dissipative losses in the order of 0.1 dB.

In previous results, as well as in those presented for the attached lines, the phase delay differs from the ideal case predicted by the transmission line

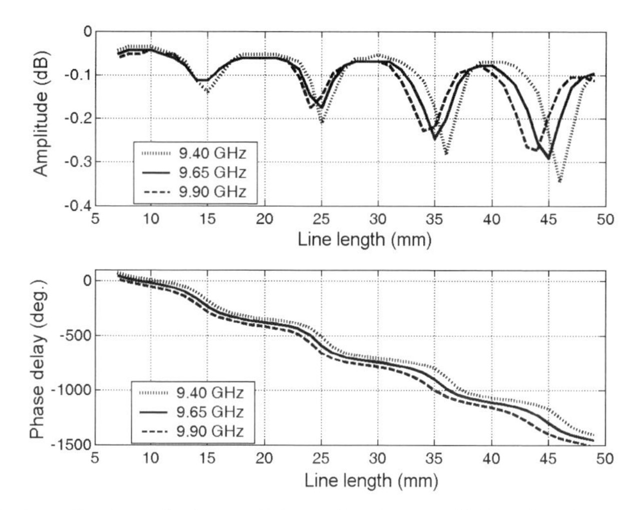

Figure 3.27. Amplitude and phase of the reflection coefficient for the periodic cell as a function of the dipole length and matching stub modified. (*E. Letters* 2006, reprinted with permission.)

TABLE 3.2. Geometrical definition of reflectarray element

Layer	ε_r	tan δ	Thickness (mm)	X (mm)	Y (mm)
D0	1.067	0.0002	7.200	—	—
Line	—	—	0.000	$W_l = 1.25$	$L_s = 1.7$
D1	3.2	0.0030	0.508	—	—
Slot	—	—	0.00	Variable 5.57–9.57	0.93
D2	1.067	0.0002	2.00	—	—
Patch	—	—	0.00	9.3	9.30
D3	3.2	0.0030	0.508	—	—

model. Then, the phase is not directly proportional to the line length, because the line resonances produce a nonlinear behavior in the phase curves. This effect can be reduced as shown in Fig. 3.26 by an appropriate redesign of the matching stubs. However, in aperture-coupled lines there is another degree of freedom, the slot length, which can be varied to adjust the coupling between line and patch, in order to compensate for the nonlinear behavior of phase response [52].

The phase delay has been computed for the a unit cell defined in Table 3.2 of period 18.5 mm × 18.5 mm, when the slot length varies from 5.57 mm to 9.57 mm and the delay line is increased from 0.00 mm to 18.49 mm, to avoid touching the neighboring delay lines without bending. The resulting phases at 10.4 GHz for normal incidence are represented as a surface in Fig. 3.28. For this case, a very linear phase curve is obtained for a slot length of 6.77 mm

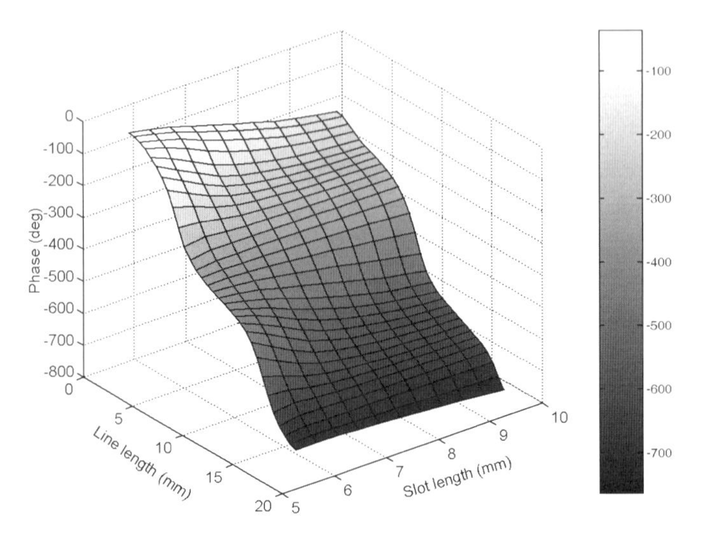

Figure 3.28. Phase shift as a function of the delay line length and the slot length.

allowing at least two cycles of phase shift delay, as shown in Fig. 3.29(a). The resulting phase response practically coincides with the ideal phase-delay ($-2\beta L$), implemented as an open-ended transmission line, which is represented with circles in the figure, as a reference. The phase curves are also represented at 9.90 GHz and 10.90 GHz, and are almost parallel, which means an excellent behavior in a 10% bandwidth. The reflection losses can be kept around 0.2 dB if a ground plane is used as shown in Fig. 3.29(b). The phase delay is also shown for oblique incidence ($\theta_i = 28°$, $\phi_i = 0°$, and 90°) in Fig. 3.30, and differences up to 60° with respect to normal incidence are observed. These results show that the angle of incidence must be taken into account for an accurate analysis, as in the case of varying-sized patches.

The phase delay range can easily be increased by using a D2 layer with a higher dielectric constant, and by bending the delay line as shown in Fig. 3.26. In conclusion, this configuration allows real time delay to be implemented in a range up to five or more 360° cycles, which can be used to design a reflectarray using true time delay in order to avoid the well known limitation in bandwidth, which will be discussed in a later chapter.

3.8 FEED MODEL AND RADIATION PATTERNS

Once the unit element has been fully characterized and a reflectarray designed, an important aspect in the reflectarray analysis is the computation of the radiation patterns. Different approaches have been used in the literature for the radiation pattern calculation. A simple method based on the integration of the

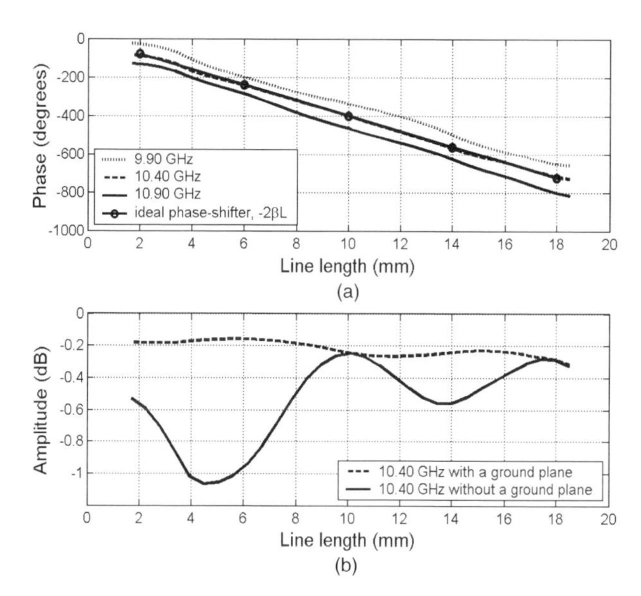

Figure 3.29. Reflection coefficient as a function of the line length for an aperture length of 6.77 mm. (a) Phase simulated compared with $-2\beta L$, (b) amplitude with and without a ground plane.

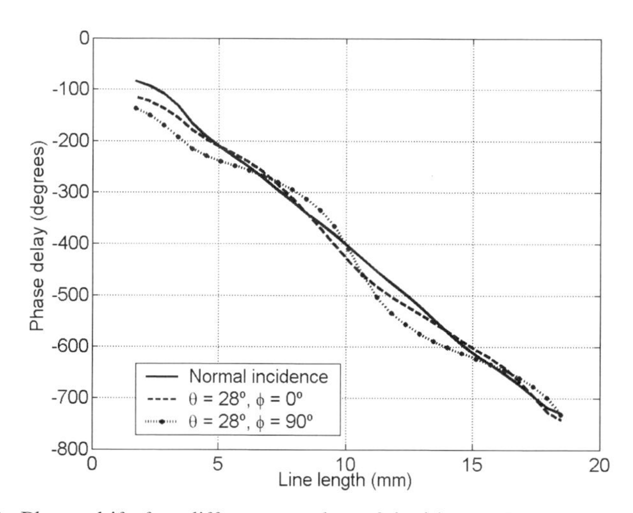

Figure 3.30. Phase-shift for different angles of incidence for an aperture-coupled element versus the line length.

aperture field on a projected aperture was proposed in [53] for a reflectarray with attached stubs. The integration in the projected aperture is reduced to a summation in all the projected reflectarray elements. This approach does not take into account the real polarization of the field, but a simple method was proposed in [53] to evaluate the cross-polarization produced by the bent lines.

For reflectarrays using varying-sized patches, the radiation pattern can be calculated from the tangential component of the reflected electric field on the reflectarray surface, as described in [54]. The formulation in [54] includes the real polarization of the field radiated by the feed, the scattered field produced by the patches including polarization coupling and the specular reflection from the ground plane.

A similar approach is described in this section, in which the radiation patterns are computed from the tangential electric field on the reflectarray elements given by Eq. (3.5), using the angular spectrum of plane waves and the stationary phase approach as described in [55]. This approach is very efficient because it computes the Fourier transform of the tangential electric field on the plane defined by the reflectarray surface ($z = 0$), assuming that the field is zero outside of the reflectarray, and it is equivalent to the second principle of equivalence. The approach ignores the residual backward field due to feed, and the diffraction at the reflectarray edges. These effects are not significant in a practical case, because the sidelobes of the horn are normally very low and the illumination at the reflectarray edges is lower than −12 dB.

The first step for the computation of the radiation patterns is to evaluate the field coming from the feed on each reflectarray element, and then the x- and y-components of the reflected field.

3.8.1 Field on the Reflectarray Elements

The field incident at each reflectarray element is determined by the position and radiation pattern of the feed. A horn is usually used as feed, and its radiation pattern is modeled as a $\cos^q(\theta)$ function in the feed coordinate system defined in Fig. 3.31. For an $\boldsymbol{x}$-polarized ideal feed, the radiated field is given by [56],

$$\boldsymbol{E}^{Fx}(\theta,\varphi) = \frac{jke^{-jkr}}{2\pi r}\left[\hat{\theta}C_E(\theta)\cos\varphi - \hat{\varphi}C_H(\theta)\sin\varphi\right], \tag{3.26}$$

and for a $\boldsymbol{y}$-polarized feed,

$$\boldsymbol{E}^{Fy}(\theta,\varphi) = \frac{jke^{-jkr}}{2\pi r}\left[\hat{\theta}C_E(\theta)\sin\varphi + \hat{\varphi}C_H(\theta)\cos\varphi\right], \tag{3.27}$$

where C_E and C_H are defined as a power of q, and determine the E- and H-plane patterns of the feed-horn

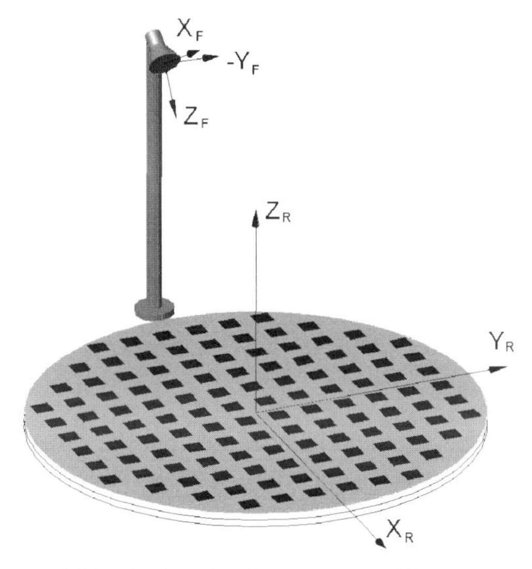

Figure 3.31. Feed and reflectarray coordinate systems.

$$C_E(\theta) = \cos^{q_E}\theta, \quad C_H(\theta) = \cos^{q_H}\theta. \tag{3.28}$$

An axial symmetric pattern is usually desirable, and the same q power is chosen in both planes, $q_E = q_H = q$.

Equations (3.26 and 3.27) give the field radiated by the feed in spherical coordinates (r, θ, φ) referred to the feed coordinate system (F), defined by the axes X_F, Y_F, Z_F (Fig. 3.31). The next step is to determine the field radiated by the feed at each reflectarray element in terms of Cartesian components in the reflectarray coordinate system (R) with axes X_R, Y_R, Z_R, also shown in Fig. 3.31. First, the field is computed at the central point of each reflectarray element from Eq. (3.26) or (3.27) depending on the polarization, then the field is transformed into Cartesian components by the expression

$$\begin{pmatrix} E_x^F \\ E_y^F \\ E_z^F \end{pmatrix} = \begin{pmatrix} \sin\theta\cos\varphi & \cos\theta\cos\varphi & -\sin\varphi \\ \sin\theta\sin\varphi & \cos\theta\sin\varphi & \cos\varphi \\ \cos\theta & -\sin\theta & 0 \end{pmatrix} \begin{pmatrix} 0 \\ E_\theta^F \\ E_\varphi^F \end{pmatrix}. \tag{3.29}$$

Note that this transformation is carried out for each element in the reflectarray, since (θ, φ) changes with position of the element. Then, the Cartesian field components are transformed from F to R coordinate system as described in [57] by the matrix equation,

$$\mathbf{E}^{RC} = \mathbf{A}\,\mathbf{E}^{FC}, \tag{3.30}$$

where matrix **A** defines the transformation of the field components from F to R coordinate system and their columns are the director cosines of the F axes on the R coordinate system. The x- and y-components of vector $\mathbf{E}^{RC}$ are the complex amplitudes (d_x, d_y) of the incident electric field on each reflectarray cell to be substituted in Eq. (3.23). From this field, the Cartesian components of the reflected field are computed as described in Section 3.5, and from them the radiation patterns. The incident field on each reflectarray element (E_θ^F, E_φ^F) can also be obtained from full-wave simulations of the horn, or from measurements of the far field radiated by the horn.

3.8.2 Radiation Patterns

Once the tangential electric field has been computed on the reflectarray elements, the field distribution is expressed in Cartesian coordinates as

$$\mathbf{E}_{RC}(x, y) = E_{Rx}(x, y)\hat{x} + E_{Ry}(x, y)\hat{y}, \tag{3.31}$$

where (x, y) are the coordinates of a point on the surface of the reflectarray in the R coordinate system. From previous aperture field distribution, the electric fields in the half-space $z > 0$ are represented by double integrals by using the angular spectrum of plane waves [55]. The radiated far field can be obtained by an asymptotic evaluation of the integrals using the stationary-phase approximation, as described in [55], giving the expression

$$\begin{aligned} \boldsymbol{E}(\theta,\varphi) = jk\Big[\left(\hat{\theta}\cos\varphi - \hat{\varphi}\sin\varphi\cos\theta\right)\tilde{E}_{Rx}(u, v) + \\ \left(\hat{\theta}\sin\varphi + \hat{\varphi}\cos\varphi\cos\theta\right)\tilde{E}_{Ry}(u, v)\Big]\frac{e^{-jk_0 r}}{2\pi r}, \end{aligned} \tag{3.32}$$

where $\tilde{E}_{Rx}(u, v)$, $\tilde{E}_{Ry}(u, v)$ are the Fourier transforms of the Cartesian components $E_{Rx}(x, y)$ and $E_{Ry}(x, y)$ of the tangential electric field, or spectral functions, defined as

$$\tilde{E}_{Rx/y}(u, v) = \iint_{RA} E_{Rx/y}(x, y)e^{jk_0(ux+vy)}dxdy, \tag{3.33}$$

where u, v are the angular coordinates, ($u = \sin\theta\cos\phi$, $v = \sin\theta\sin\phi$).

The double integral in (3.33) must be extended to the entire plane $z = 0$, but is limited to the reflectarray surface, because the tangential electric field is assumed to be zero outside of the reflectarray. This assumption is equivalent to the second principle of equivalence, in which the antenna aperture is assumed to be immersed in a perfect conductor plane. The same expression (3.32) is obtained if the far field is computed from the equivalent currents assuming the second principle of equivalence [58].

In order to evaluate the integral (3.33) element by element, the following change of variables is defined for the coordinates (x, y):

$$\begin{aligned} x &= x' + mp_x - \frac{(N_x - 1)p_x}{2}; \quad m = 0, 1, 2, \ldots, N_x - 1 \\ y &= y' + np_y - \frac{(N_y - 1)p_y}{2}; \quad n = 0, 1, 2, \ldots, N_y - 1, \end{aligned} \tag{3.34}$$

where the coordinates of the central point of the element (m, n) are $\left(mp_x - \frac{(N_x - 1)p_x}{2}, np_y - \frac{(N_y - 1)p_y}{2}\right)$; and x', y' are limited to a periodic cell $\left(-\frac{p_x}{2} \leq x' < \frac{p_x}{2}, -\frac{p_y}{2} \leq y' < \frac{p_y}{2}\right)$. N_x and N_x are the maximum number of elements in the x- and y-direction, respectively. Substituting (3.34) into (3.33), the spectral functions for x/y components are rewritten as

$$\tilde{E}_{Rx/y}(u, v) = K_1 \cdot \sum_{m=0}^{N_x-1} \sum_{n=0}^{N_y-1} \left[e^{jk_0(ump_x + vnp_y)} \int_{-\frac{p_x}{2}}^{\frac{p_x}{2}} \int_{-\frac{p_y}{2}}^{\frac{p_y}{2}} E_{Rx/y}^{m,n}(x', y') e^{jk_0(ux' + vy')} dx'dy' \right], \tag{3.35}$$

where

$$K_1 = e^{-j\frac{k_0}{2}[u(N_x-1)d_x + v(N_y-1)d_y]}. \tag{3.36}$$

Considering that the tangential field components in each reflectarray cell are directly the complex coefficients of the reflected field a_x and a_y in Eq. (3.24), the amplitude and phase are uniform in each cell (m, n),

$$E_{Rx/y}^{m,n}(x', y') = a_{x/y}(m, n) = A_{x/y}(m, n) e^{j\phi_{x/y}(m,n)}, \tag{3.37}$$

where $A_{x/y}(m, n)$ and $\phi_{x/y}(m, n)$ are, respectively, the amplitude and phase of the complex coefficients $a_{x/y}(m, n)$. The spectral function (3.35) is written as a summation after carrying out the integration on the periodic cells,

$$\begin{aligned} \tilde{E}_{Rx/y}(u, v) &= K_1 p_x p_y \operatorname{sinc}\left(\frac{k_0 u p_x}{2}\right) \\ &\operatorname{sinc}\left(\frac{k_0 v p_y}{2}\right) \cdot \sum_{m=0}^{N_x-1} \sum_{n=0}^{N_y-1} A_{x/y}(m, n) e^{j\phi_{x/y}(m,n)} e^{jk_0(umd_x + vmd_y)}. \end{aligned} \tag{3.38}$$

Note that the summation is extended to a rectangular grid on N_x by N_y elements, but if the reflectarray is circular or elliptical, the amplitude of the field for the elements outside of the reflectarray will be set to zero. The radiation pattern is obtained by substituting the spectral functions $\tilde{E}_{Rx}(u, v)$ and $\tilde{E}_{Rx}(u, v)$ in (3.32).

The double summation in (3.38) can be easily identified with a 2-D Inverse Discrete Fourier Transform (IDFT2) defined as,

$$f(p,q) = IDFT2[F(m,n)] = \frac{1}{MN}\sum_{m=0}^{M-1}\sum_{n=0}^{N-1} F(m,n)e^{j\frac{2mp\pi}{M}}e^{j\frac{2nq\pi}{N}}. \tag{3.39}$$

Then, the spectral functions and the radiation patterns can be computed very efficiently by using a 2-D FFT algorithm. The limits in the summations (M, N) in principle are equal to the elements in the grid defined in the reflectarray (N_x, N_y), to identify (3.38) and (3.39). When applying IDFT2, the spectral functions are obtained in a discrete number of angular coordinates (u, v) given by:

$$u = \frac{2\pi}{Mp_x k_0}p; \qquad p = 0, 1, 2, \cdots M-1 \tag{3.40}$$

$$v = \frac{2\pi}{Np_y k_0}q; \qquad q = 0, 1, 2, \cdots N-1. \tag{3.41}$$

After a shift in the variables (p, q) and (m, n), the range in u,v is made symmetric,

$$-\lambda_0/2p_x \le u \le \lambda_0/2p_x; \qquad -\lambda_0/2p_y \le v \le \lambda_0/2p_y. \tag{3.42}$$

Note that if the periodic cell is $\lambda_0/2$, the radiation patterns will be computed in the whole visible range, defined by the circle $u^2 + v^2 \le 1$. If the period is larger, the radiation patterns will be obtained for a smaller region of (u, v), unless the periodic cell is subdivided in four subcells for the summations in (3.38). Normally, the period is chosen between 0.5λ and 0.7λ to avoid grating lobes, and the range (u, v) is chosen to be enough to represent the range of interest for the radiation patterns.

When applying IDFT2, the radiation pattern is evaluated in a number of (u, v) points equal to the number of elements defined on the reflectarray surface ($N_x \times N_x$). However, the radiation patterns can be computed in a larger number of (u, v) points by extending the grid on the reflectarray plane ($M > N_x$, $N > N_y$) and setting the amplitude equal to zero for all the elements outside of the reflectarray.

Once the spectral functions $\tilde{E}_{Rx}(u,v)$ and $\tilde{E}_{Ry}(u,v)$ are computed and substituted in (3.32), the radiation pattern is obtained in terms of (θ, φ) components,

$$E_\theta(\theta,\varphi) = jk\frac{e^{-jkr}}{2\pi r}(\tilde{E}_{Rx}\cos\varphi + \tilde{E}_{Ry}\sin\varphi) \tag{3.43}$$

$$E_\varphi(\theta,\varphi) = -jk\frac{e^{-jkr}}{2\pi r}\cos\theta\,(\tilde{E}_{Rx}\sin\varphi - \tilde{E}_{Ry}\cos\varphi). \tag{3.44}$$

However, it is more common to use the co-polar and cross-polar components of the field according to the third Ludwig definition [59]. For an x-polarized feed horn, the transformation from (θ, φ) to co-polar (E_p) and cross-polar (E_q) components is given by

$$\begin{pmatrix} E_p^X \\ E_q^X \end{pmatrix} = \begin{pmatrix} \cos\varphi & -\sin\varphi \\ -\sin\varphi & -\cos\varphi \end{pmatrix} \begin{pmatrix} E_\theta \\ E_\varphi \end{pmatrix}, \tag{3.45}$$

and for a y-polarized feed by

$$\begin{pmatrix} E_p^Y \\ E_q^Y \end{pmatrix} = \begin{pmatrix} \sin\varphi & \cos\varphi \\ \cos\varphi & -\sin\varphi \end{pmatrix} \begin{pmatrix} E_\theta \\ E_\varphi \end{pmatrix}. \tag{3.46}$$

The technique described in this section for the computation of the radiation patterns allows an accurate prediction of the co- and cross-polar radiation patterns. This includes all possible sources of cross-polarization, such as from the horn, the geometrical projections and the reflectarray elements. For example, if an ideal x-polarized feed-horn is considered, the (θ, φ) components of the field are given by (3.26) and after transformation into Cartesian components on the reflectarray elements, Eqs. (3.29 and 3.30) gives a small component of E_y out of the symmetry planes produced by the geometrical projection of the vectors. Additional cross-polarization is introduced when multiplying by the reflection coefficients, as a result of the polarization coupling at the reflectarray elements, as seen from Eq. (3.21 and 3.24). Therefore, for an ideal linear polarized horn, the field distribution on the reflectarray surface has both x- and y-components, as in (3.31), and the co-polar and cross-polar components of the field are obtained from them by (3.45). The cross-polarization produced by the feed can also be included if the real radiation patterns of the horn are used instead of (3.26) or (3.27).

One approximation used in the technique described is the assumption that the tangential electric field is zero outside of the reflectarray, which implies the use of the second principle of equivalence, in which the antenna aperture is assumed to be immersed in a perfect conductor plane. The approach ignores the diffraction at the reflectarray edges, which should not be significant, because of the low illumination level at the edges. However, the radiation patterns can be computed in a more accurate way by using both tangential electric and magnetic fields on the reflectarray surface, and the first principle of equivalence. The tangential magnetic field can be obtained from the complex amplitudes a_l of reflected electric field as shown in (3.18), because the tangential magnetic and electric modal fields, $\mathbf{h}_l$ and $\mathbf{e}_l$, are related by the TE and TM wave impedances.

Another approximation considered in this section is the assumption that the field is uniform in each periodic cell, being obtained as the combination of only the fundamental TE and TM Floquet harmonics, which means that the higher-order harmonics used in the analysis of the cell are ignored in the computation of the radiation patterns. This approximation can also be eliminated, if the field at each reflectarray cell is computed as a summation of all the Floquet harmonics as in (3.5) and (3.18). Then, the fields are not uniform on each cell, and Eq. (3.37) must be substituted by a summation of Floquet harmonics, which after the integration in the periodic cell will result in a summation of sinc functions in (3.38), involving the complex amplitudes of the higher order modes. However, the effect of the evanescent Floquet harmonics in the radiation patterns is limited to variations in the side lobes, as shown in [60].

3.8.3 Experimental Results

A circular two-layer reflectarray based on the periodic structure of Fig. 3.13 of 406 mm diameter has been designed, built and measured, see Fig. 3.32. The reflectarray was designed to radiate a pencil beam in the direction $\theta_b = 25°$, $\varphi_b = 0°$ at 11.95 GHz for both linear polarizations. A commercial horn was

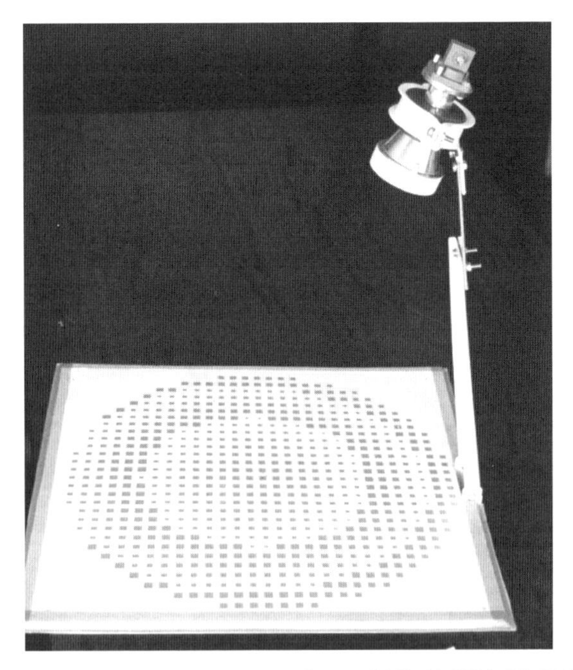

Figure 3.32. Built prototype. (From [15], © 2001 IEEE.)

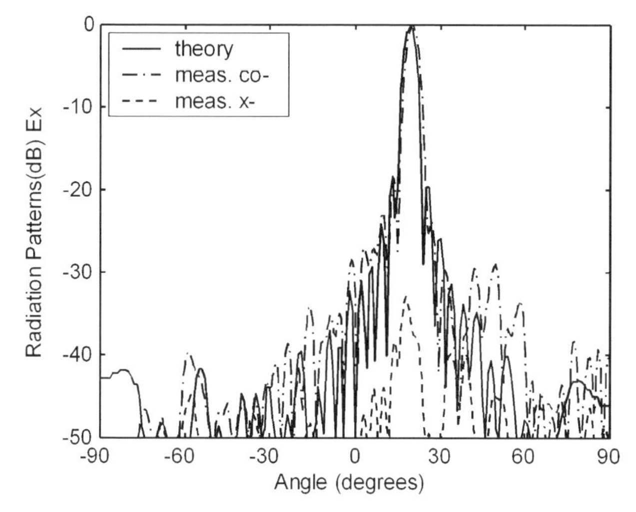

Figure 3.33. Comparison between theoretical and measured patterns. (From [15], © 2001 IEEE.)

used with its phase center located at coordinates $X_f = -150$, $Y_f = 0$, $Z_f = 300$ mm in the reflectarray coordinate system. The radiation patterns have been measured and computed by the technique described in this section. Fig. 3.33 shows the patterns on the *x-z* plane at 11.95 GHz for *x* polarization, illustrating a good agreement between measured and theoretical patterns. The results obtained for orthogonal polarization are very similar.

3.8.4 Gain Computation

The directivity can be easily computed from the far field radiated by the reflectarray. The antenna gain can also be computed using the input power in the feed-horn P_F, according to

$$G(\theta, \varphi) = \frac{|\mathbf{E}(\theta, \varphi)|^2 \, 4\pi r^2}{2\eta_0 P_F}, \tag{3.47}$$

where $\eta_0 = \sqrt{\mu_0 / \varepsilon_0}$ is the intrinsic impedance of the free-space. Assuming the ohmic losses in the feed-horn are negligible, typically they are in the order of 0.05 dB, P_F can be computed as the total power radiated by the horn as

$$P_F = \int_{\theta=0}^{\pi/2} \int_{\varphi=0}^{2\pi} \frac{|\mathbf{E}^F(\theta, \varphi)|^2}{2\eta_0} r^2 \sin\theta \, d\theta \, d\varphi, \tag{3.48}$$

where $\mathbf{E}^{\mathrm{F}}(\theta, \varphi)$ is the field radiated by the horn given by Eqs. (3.26) or (3.27) for x- or y-polarization, respectively. The amplitude squared for both linear polarizations is

$$|\mathbf{E}_x^{\mathrm{F}}|^2 = |\mathbf{E}_y^{\mathrm{F}}|^2 = \frac{k_0^2 \cos^{2q}\theta}{4\pi^2 r^2}, \tag{3.49}$$

and substituting (3.49) in (3.48), the total power radiated by the feed is

$$P_F = \frac{\pi}{\eta_0 \lambda^2 (2q+1)}. \tag{3.50}$$

Since the antenna requirements are usually given in co-polar gain, the co-polar and cross-polar patterns are expressed in gain, using the following expressions:

$$G_p(\theta, \varphi) = \frac{|E_p(\theta, \varphi)|^2 \, 4\pi r^2}{2\eta_0 P_F}; \qquad G_q(\theta, \varphi) = \frac{|E_q(\theta, \varphi)|^2 \, 4\pi r^2}{2\eta_0 P_F}. \tag{3.51}$$

The co-polar gain computed by (3.51) includes the spillover losses and the ohmic losses in the reflectarray, if they were included in the analysis of the reflectarray elements. As previously mentioned, the field radiated by the horn obtained from measurements or from full-wave simulations can be used instead of (3.26) or (3.27) for the computations of the radiation patterns. In that case, $\mathrm{P_F}$ will be substituted by the power radiated by the horn.

REFERENCES

1. R. E. Munson, H. A. Haddad, and J. W. Hanlen, "Microstrip reflectarray for satellite communications and RCS enhancement or reduction," *US patent* 4684952, Aug. 1987.
2. J. Huang, "Microstrip reflectarray," *IEEE Intl. Symp. Antennas Propagat.*, Ontario, Canada, pp. 612–615, June 1991.
3. D.-C. Chang and M.-C. Huang, "Microstrip reflectarray antenna with offset feed." *Electr. Lett.*, pp. 1489–1491, July 1992.
4. R. D. Javor, X.-D. Wu, and K. Chang. "Design and performance of a microstrip reflectarray antenna." *Microwave and Optical Technology Letters*, Vol. 7, No. 7, pp. 322–324, May 1994.
5. D. G. Gonzalez, G. E. Pollon, and J. F. Walker, "Microwave phasing structures for electromagnetically emulating reflective surfaces and focusing elements of selected geometry," Patent US 4905014, Feb. 1990.
6. D. M. Pozar and T. A. Metzler, "Analysis of a reflectarray antenna using microstrip patches of variable size," *Electr. Lett.* Vol. 29, No. 8, pp. 657–658, April 1993.

7. S. D. Targonski and D. M. Pozar. "Analysis and design of a microstrip reflectarray using patches of variable size," in Antennas and Propagation Society International Symposium, 1994. AP-S. Digest, pp. 1820–1823, June 1994.
8. D. M. Pozar and S. D. Targonski, "A microstrip reflectarray using crossed dipoles," *IEEE Antennas and Propagation Society International Symposium*, pp. 1008–1011, June 1998.
9. A. W. Robinson, M. E. Bialkowski, and H. J. Song, "An X-band passive reflect-array using dual-feed aperture-coupled patch antennas," *Asia Pacific Microwave Conference*, pp. 906–909, December 1999.
10. M. G. Keller, M. Cuhaci, J. Shaker, and A. Petosa, A. Ittipiboon, and Y. M. M. Antar, "Investigations Novel Reflectarray Configurations," *Symposium on Antenna Technology and Applied Electromagnetics*, pp. 299–302, 2000.
11. M. R. Chaharmir, J. Shaker, M. Cuhaci, and A. Sebak, "Reflectarray with variable slots on ground plane," *IEE Proc.-Microw. Antennas Propagat.*, Vol. 150, No. 6, pp. 436–439. December 2003.
12. J. Huang, "A Ka-band microstrip reflectarray with elements having variable rotation angles," *IEEE Trans. Antennas Propagat.*, Vol. 46, No. 5, pp. 650–656, May 1998.
13. A. Martynyuk, J. I. Martínez, and N. Martynyuk, "Spiraphase-type reflectarrays based on loaded ring slot resonators," *IEEE Trans. Antennas Propagat.*, Vol. 52, No. 1, pp. 142–153, January 2004.
14. J. A. Encinar, "Printed circuit technology multilayer planar reflector and method for the design thereof," European Patent EP 1120856, June 1999.
15. J. A. Encinar, "Design of two-layer printed reflectarrays using patches of variable size," *IEEE Trans. Antennas Propagat.*, Vol. 49, No. 10, pp. 1403–14010, October 2001.
16. J. A. Encinar and J. A. Zornoza, "Broadband design of three-layer printed reflectarrays," *IEEE Trans. Antennas Propagat.*, Vol. 51, no. 7, pp. 1662–1664, July 2003.
17. Y. Zhuang, C. Wu, K.-L. Wu, and J. Litva, "Microstrip reflectarrays: Full-wave analysis and design scheme." In Antennas and Propagation Society International Symposium, 1993. AP-S. Digest, pages 1386–1389, July 1993.
18. Y. Zhuang, J. Litva, C. Wu, and K.-L. Wu, "Modelling studies of microstrip reflectarrays," *IEE Proc. Microw Propag.*, Vol. 142, No. 1, pp. 78–80, Feb. 1995.
19. D. M. Pozar, "Microstrip reflectarrays myths and realities," JINA 2004, International Symposium on Antennas, Nice, France, pp. 175–179, November 2004.
20. T. Metzler and D. Shaubert, "Scattering from a stub loaded microstrip antenna," Antennas and Propagation Society International Symposium, 1989. AP-S. Digest, pp. 446–449, June 1989.
21. F. Venneri, G. Angiulli, and G. Di Massa, "Design of microstrip reflectarray using data from isolated patch," *Microwave and Optical Technology Letters*, Vol. 34, pp. 411–414, Sept. 2002.
22. R. P. Jedlicka, M. T. Poe, and K. R. Carver, "Measured mutual coupling between microstrip antennas," *IEEE Trans. Antennas Propagat.*, Vol. 29, pp. 147–149, Jan. 1981.
23. D. Cadoret, A. Laisne, M. Milon, R. Gillard, and H. Legay, "FDTD analysis of reflectarray radiating cells," IEEE/ACES International Conference on Wireless

Communications and Applied Computational Electromagnetics, 2005, pp. 853–856, April 2005.

24. D. M. Pozar and D. H. Schaubert "Analysis of an infinite array of rectangular microstrip patches with idealized probe feeds," *IEEE Trans. on Antennas Propagat.*, Vol. 32, pp. 1101–1107, Oct. 1984.
25. D. M. Pozar, "Analysis of an infinite phased array of aperture coupled microstrip patches," *IEEE Trans. on Antennas Propagat.*, Vol. 37, pp. 418–425, April 1989.
26. R. Mittra, C. H. Chan, and T. Cwik, "Techniques for analyzing frequency selective surfaces a review," *Proc. IEEE*, Vol. 76, no. 12, pp. 1593–1615, Dec. 1988.
27. C. Wan and J. A. Encinar, "Efficient computation of Generalized Scattering Matrix for analyzing multilayered periodic structures," *IEEE Trans. on Antennas Propagat.*, Vol. 43, pp. 1233–1242, Oct. 1995.
28. I. Bardi, R. Remski, D. Perry, and Z. Cendes, " Plane wave scattering from frequency selective surfaces by finite element method," *IEEE Trans. Magn.* Vol. 38 no. 2, pp. 641–644, March 2002.
29. P. Harms, R. Mittra, and K. Wae, "Implementation of periodic boundary condition in finite-difference time-domain algorithm for FSS structures," *IEEE Trans. on Antennas Propagat.*, Vol. 42, pp. 1317–1324, Sept. 1994.
30. M. Lambea, M. A. González, J. A. Encinar, and J. Zapata, "Analysis of frequency selective surfaces with arbitrarely shaped apertures by Finite Element Method and Generalized Scattering Matriz," 1995 IEEE Antennas and Propagation Intl. Symposium, Newport Beach, California, USA, pp. 1644–1647.
31. E. Guirard, R. Moulinet, and R. Gillard, "An FDTD optimization of a circularly polarized reflectarray unit cell," IEEE Antennas and Propagation Soc. Intl Symposium, San Antonio, Texas, Vol. 3, pp. 136–139, June 2002.
32. M. G. Keller, J. Shaker, A. Petosa, A. Ittipibon, M. Cuhaci, and Y. M. M. Antar, "A Ka-band dielectric resonator antenna reflectarray," European Microwave Conference 2000, Paris, France, pp. 272–275, Oct. 2000.
33. J. A. Encinar, L. Datashvili, J. Agustín Zornoza, M. Arrebola, M. Sierra-Castañer, J. L. Besada, H. Baier, and H. Legay, "Dual-polarization dual-coverage reflectarray for space applications," *IEEE Trans. on Antennas Propagat.*, vol. 54, pp. 2827–2837, Oct. 2006.
34. M. Bozzi and L. Perregrini, "Analysis of multilayered printed frequency selective surfaces by the MoM/BI-RME method," *IEEE Trans. Antennas Propagat.*, Vol. 51, pp. 2830–2836, Oct. 2003.
35. M. Bozzi, S. Germani, and L. Perregrini, "Performance comparison of different element shapes used in printed reflectarrays", *IEEE Antennas and Wireless Propagation Letters*, Vol. 2, pp. 219–222, 2003.
36. D. Pilz and W. Menzel, "Full wave analysis of a planar reflector antenna," 1997 Asia Pacific Microwave Conference, pp. 225–227, Dec. 1997.
37. K. F. Chan, K. W. Lam, C. H. Chan, and R. Mittra, "Fast computational techniques for the analysis of microstrip reflectarrays," private communication.
38. I. Rullhusen and F. Arndt, "Efficient accurate analysis of reflectarrays using the adaptive integral method," 12e Journées Internationales de Nice sur les Antennes (JINA), Nice, France, pp. 367–370, Nov. 2002.

39. F. Venneri, L. Boccia, G. Angiulli, G. Amendola, and G. Di Massa, "Analysis and design of passive and active reflectarrays," *Int. Journal of RF and Microwave Computer-Aided Engineering*, pp. 370–377, August 2003.
40. F.-C.E. Tsai and M. E. Bialkowski, "Designing a 161-element Ku-band microstrip reflectarray of variable size patches using an equivalent unit cell waveguide approach," *Antennas and Propagation, IEEE Transactions on*, Vol. 51, Issue 10, Part 2, pp. 2953–2962, Oct. 2003.
41. J. A. Encinar and R. Caballero, "Design and development of a double-sandwich dichroic subreflector," 24th Antenna Workshop on Innovative Periodic Antennas: Photonic Bandgap, Fractal and Frequency Selective Surfaces, ESTEC, Noordwijk, The Netherlands, pp. 115–119, 30 May– 1 June 2001.
42. R. Caballero, C. Palacios, and J. A. Encinar, "Marx Express and Venus Express High Gain Antennas," 27th ESA Antenna workshop on Innovative Periodic Antennas: Electromagnetic Bandgap, Left-Handed Materials, Fractal and Frequency Selective Surfaces, Santiago de Compostela, Spain, pp. 83–89, 9–11 March 2004.
43. P. Gay-Balmaz, J. A. Encinar, and J. R. Mosig, "Analysis of a multilayer printed arrays by a modular approach based on the generalized scattering matrix," *IEEE Trans. Antennas Propagat.*, Vol. 48, No. 1, pp. 26–34, January 2000.
44. J. R. James, P. S. Hall, and C. Wood, *Microstrip Antenna Theory and Design*, Peter Peregrinus Ltd., Stevanage, UK, pp. 100, 1981.
45. M. Lambea and J. A. Encinar, "Analysis of multilayer frequency selective surfaces with rectangular geometries," Ninth Int. Conf. on Antennas and Propagation ICAP'95, Eindhoven, The Netherlands, pp. 528–531, 1995.
46. P. W. Hannan and M. A. Balfour, "Simulation of phased array antennas in waveguides," *IEEE Trans. on Antennas Propagat.*, Vol. 13, pp. 342–353, May 1965.
47. A. W. Robinson, M. E. Bialkowski, and H. J. Song, "An 137-element active reflectarray with dual-feed microstrip patch elements," *Microwave and Optical Technology Letters*, Vol. 26, No. 3, pp. 147–151, August 2000.
48. J. R. James and P. S. Hall, *Handbook of Microstrip Antennas*, Peter Peregrinus Ltd. pp. 330–337, 1989.
49. D. M. Pozar, "Microstrip antennas," *Proc. of IEEE*, Vol. 80, pp. 79–91, Jan. 1992.
50. E. Carrasco, B. Alfageme, and J. A. Encinar, "Design of a multilayer aperture-coupled cell used as phase-shifter in reflectarrays," *Journées Internationales de Nice sur les Antennes*, Nice, France, November 2004.
51. E. Carrasco, M. Barba, and J. A. Encinar, "Aperture-coupled reflectarray element with wide range of phase delay," *Electronics Letters 8th*, Vol. 42, pp. 667–668, No. 12, June 2006.
52. E. Carrasco, M. Barba, and J. A. Encinar, "Reflectarray element based on aperture-coupled patches with slots and lines of variable length," *IEEE Trans. on Antennas and Propagat.*, Vol. 55, pp. 820–825, March 2007.
53. D.-C. Chang and M.-C. Huang, "Multiple-polarization microstrip reflectarray antenna with high efficiency and low cross-polarization," *IEEE Trans. Antennas and Propagation*, vol. 43, no. 8, pp. 829–834, August 1995.
54. D. M. Pozar, S. D. Targonski, and H. D. Syrigos, "Design of millimeter wave microstrip reflectarray," *IEEE Trans. Antennas and Propagation*, Vol. 45, pp. 287–295, Feb. 1997.

55. R. H. Clark and J. Brown, *Difraction Theory and Antennas*. Chichester: Ellis Horwood, pp. 85–86, 1980.
56. Y. T. Lo and S. W. Lee, *Antenna Handbook. Vol. I: Antenna Fundamentals and Mathematical Techniques*. Van Nostrand Reinhold, pp. 1.28–29, 1993.
57. Y. Rahmat-Samii. "Useful coordinate transformations for antenna applications," *Antennas and Propagation, IEEE Trans. on*; pp. 571–574. July 1979.
58. W. L. Stuzman and G. A. Thiele, *Antenna Theory and Design*, 2nd edition John Wiley and Sons, pp. 275–283, 1988.
59. A. C. Ludwig, "The definition of cross polarization," *IEEE Trans. Antennas Propagat.*, pp. 116–119, Jan. 1973.
60. J. A. Encinar, "Analysis and CAD techniques for periodic leaky-wave printed antennas: Numerical and experimental results," *International Journal of Microwave and Millimeter-Wave Computer-Aided Engineering*, Vol. 4, No.1, Jan. 1994, pp. 88–99.

CHAPTER 4

Practical Design Approach

The previous chapter gives detailed descriptions of various techniques for analyzing and deriving the desired phase as a function of element change, which is the most important step in achieving an efficient reflectarray. Other areas that are also important in leading to a good design are knowing the element effects and its selection criteria, calculating the path length difference and the required phase delay, calculating the far-field pattern, and knowing how to design an efficient reflectarray geometry. These important practical approaches in designing an efficient reflectarray are discussed separately below.

4.1 ELEMENT EFFECTS AND SELECTION

The most important and critical step of the reflectarray design is its element characterization. If the element design is not optimized, the reflectarray will not scatter the signal from the feed effectively to form an efficient far-field beam. Four important parameters that govern the element selection and design are the element reflection phase, the element beamwidth, the element bandwidth, and the element reflection efficiency. These are discussed separately below.

4.1.1 Element Reflection Phase

The previous chapter presented detailed techniques for characterizing reflectarray elements' reflection phases. Here, it is briefly discussed again. The phase change versus element change (patch size, delay line length, etc.) must be calibrated correctly. One of the most popular techniques for calibrating the phase is to use the infinite-array approach [1, 2] to include local mutual coupling effect due to surrounding elements.

It is not feasible for the current computer technology to have a complete, rigorous solution to include all the mutual coupling effect of all different

Reflectarray Antennas, by John Huang and José A. Encinar

elements since the reflectarray generally consists of too many elements. This infinite-array approach, which assumes all surrounding elements are identical, is a periodic solution using Floquet modes. It is an approximation, but, in many cases, is adequate enough. This infinite array approach can be carried out by using the Method of Moments technique [1, 2] or equivalently done by a Finite-Difference Time-Domain (FDTD) analysis on a unit cell of single element [3]. Mathematical waveguide simulator, which simulates the infinite array approach, can also be adapted by using the commercial software, HFSS (a finite element technique), to achieve the element phase information.

Fig. 4.1 shows the very popular approach of using an H-wall waveguide simulator, also called parallel-plate waveguide simulator, where the top and bottom surfaces of the waveguide are electric conducting walls, while the right and left walls are magnetic field walls. The vertically polarized incoming waves will see the element at the end of the waveguide at the broadside direction and then scattered back also at the broadside direction with a set of amplitude and phase information. For a circularly polarized element, a second set of amplitude and phase information should be recorded with the incoming wave having a 90° phase shift and the element rotated 90°. This process should be repeated for different element changes until a 360° phase-change range can be obtained. This H-wall waveguide simulator, however, can only calibrate the reflectarray element for the normal incident case. Although this is adequate, more accurate results could be achieved by calibrating the element with various incident angles. To accomplish this, a real physical waveguide model can be constructed with the element placed at the end of the waveguide. It is known that the incident wave is composed of two plane waves traveling at an off-broadside angle (θi) that obeys the relationship of $\sin(\theta i) = \pi/(ka)$, where "a" is the H-plane or generally the long dimension of the waveguide. This waveguide measurement approach, due to the image theory, approximates an infinite array scenario where the elements are separated by a distance of "a" along the H-plane and a distance of "b" along the E-plane. Each waveguide

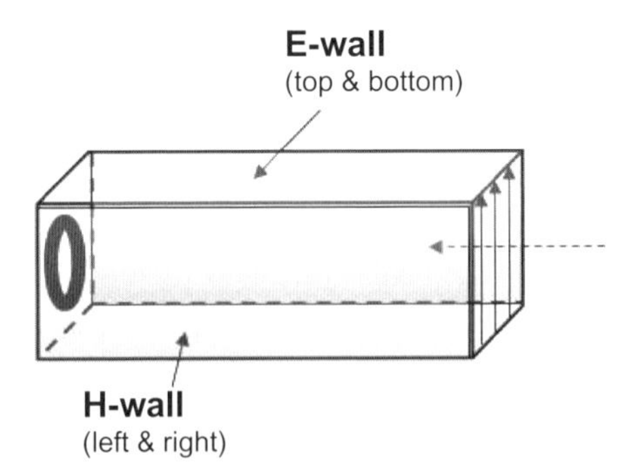

Figure 4.1. An H-wall waveguide simulator to calibrate the reflectarray element phase-change versus element-change.

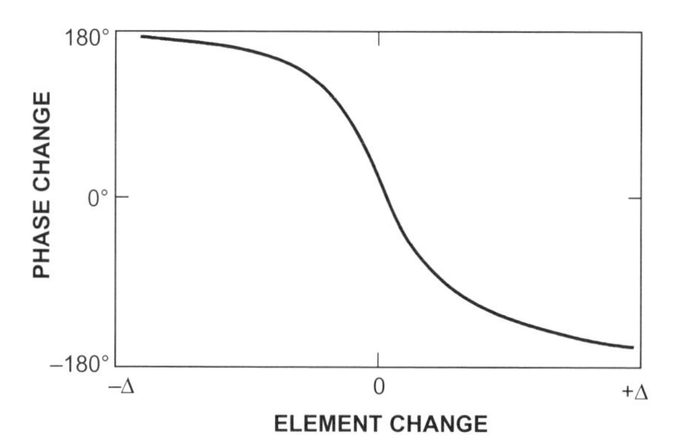

Figure 4.2. A typical "S" curve of a reflectarray element phase-change versus element-change.

can only simulate one incident angle at one frequency. Several different waveguides need to be constructed in order to calibrate the element with several sets of calibration data for several different incident angles. By changing the frequency, a slight variation of incident angle can also be achieved.

All of the above techniques are used to derive the phase-versus-element-change curve, which is generally an S-shaped curve with nonlinear relationship as illustrated in Fig. 4.2. The antenna designer should minimize the slope at the center of the curve so that the phase change will not be overly sensitive to the element change. If the curve is too steep, the element change or fabrication tolerance may become an issue, in particular at high microwave frequencies. It is also important to make sure that the element selected will have enough element variation to yield at least a phase change of ±180°.

4.1.2 Element Beamwidth

In addition to its phase, the element's beamwidth is also a crucially important design parameter. Its beamwidth must correlate correctly with the reflectarray's f/D ratio to accommodate most incident angles from the feed. As an example, if the reflectarray element has a relatively narrow beamwidth (by using Vivaldi radiator, horn, or patch with thick substrate) and a small f/D ratio as illustrated in Fig. 4.3, the elements located near the edge of the reflectarray will not receive energy effectively from the illuminating feed. On the other hand, if the element has too wide a beam with lower gain, the overall reflectarray may suffer with a slightly lower antenna gain. Although the array gain is mostly dictated by its aperture size, the element gain of the array does impact somewhat the overall array gain. It would be ideal if the elements close to the center have narrow beamwidth with higher gain, while the elements close to the edge have wider beamwidth. However, this may make manufacturing of the antenna more difficult and more costly.

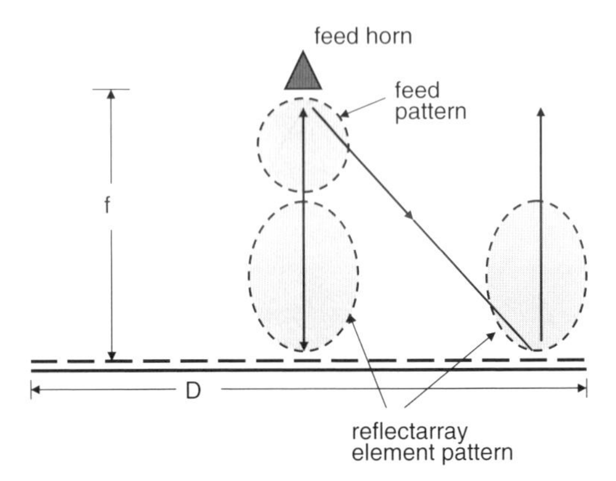

Figure 4.3. Reflectarray element pattern effect.

4.1.3 Element Bandwidth

One of the factors that limits the bandwidth of a reflectarray is its element bandwidth. The microstrip patch element generally has a bandwidth of about 3 to 5 percent. To achieve wider bandwidth for a printed patch element, techniques such as using stacking multiple patches [4, 5] and using sequentially rotated subarray elements [6] have been employed. More than 15 percent bandwidths have been reported. More detailed discussion on reflectarray bandwidth will be given in Chapter 5.

4.1.4 Element Reflection Efficiency

When a wave is incident upon a reflectarray patch element, all energy must be scattered back due to the existence of the ground plane. There will be three back-scattered components. One is the reradiated component due to the resonant activity of the patch, the second is the specular reflected component due to the ground plane, and the third is the scattered component due to the nonresonant structures of the patch and any attached delay line. Because the patch elements generally have thin substrates with thickness much less than 0.1 λ_o, only the first two components dominate. For linear or dual-linear polarization, there are two types of reflectarray elements that can be used, as indicated in Fig. 1.2 (a), (b), and (c) of Chapter 1. Fig. 1.2 (a) uses identical patch sizes with variable-length phase delay lines.

If one analyzes the radar cross-sections (RCS) of the element of a reflectarray that uses identical patch sizes with variable-length delay line attached, one will find out that, although all patches have the same size, the maximum RCS of each element will occur at different frequencies for different phase delay line lengths. In other words, the attached phase delay line becomes part of the resonating structure of the reflectarray element and, thus, cause the identical-size patches to resonate at different frequencies. For the elements of Fig. 1.2

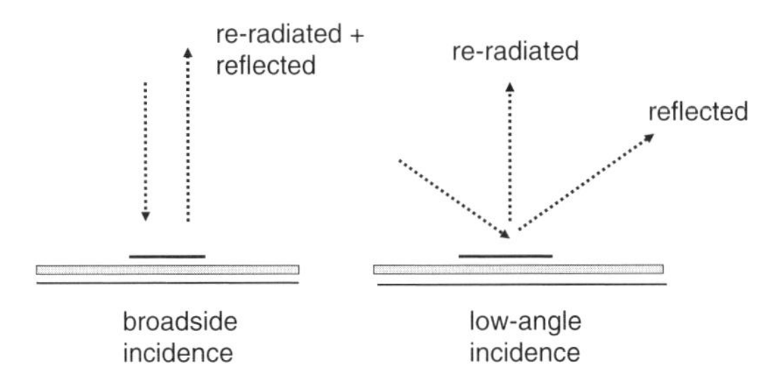

Figure 4.4. Reradiated and reflected waves of a reflectarray element. Right figure showing low incident angle wave for an edge element.

(b) and (c) with variable patch sizes, different-size elements will certainly also resonate at different frequencies. Consequently, regardless of the type of elements used, at the design frequency (f_o), only the elements having resonant dimension will have the complete re-radiated energy, while rest of the elements will back-scatter with both reradiated and reflected energy.

As an example illustrated in Fig. 4.4 for a center-fed, broadside-radiated reflectarray, only the elements located within the center region of the reflectarray will have both reradiated and reflected waves scattered back in the same direction. For those elements located close to the edge of the reflectarray with large incident angles from the feed's incident rays, the reradiated component will be directed toward the main beam, while the reflected component will be directed away from the main beam. It is these edge elements' reflection components that contribute to the sidelobes and the inefficiency of the reflectarray. To minimize this inefficiency, one could design the reflectarray with large f/D ratio and with elements having minimized reflection component (discussed further in the next section). It should be pointed out here that, even though when designing the reflectarray the phase adjustment is based on the total field of the reradiated and the reflected components (see Section 3.5 of Chapter 3), most of the reflected energy, if its direction does not coincide with the main beam direction, is wasted.

For circular polarization, if one selects the variable-angular-rotation approach as indicated in Fig. 1.2(d), the above-mentioned inefficiency would be eliminated. Because now all elements are truly identical and all elements will resonate at the same design frequency, there will be very little specular reflected component regardless of the incident angle of the waves.

4.1.5 Element Spacing

All elements in the reflectarray must have proper spacing between adjacent elements in order to avoid the grating lobe type of radiation. For broadside incident and radiated waves, the element spacing could be relatively large (e.g., $>0.9\,\lambda_o$). For elements located toward the edge, the wave's incident angles from the feed horn could be large, and the element spacing requirement will become smaller and should be governed by the following conventional array equation:

$$\frac{d}{\lambda_O} \leq \frac{1}{1+\sin\theta}, \tag{4.1}$$

where d is element spacing and θ is either the incident angle from the feed or the main beam tilt angle from the broadside direction, whichever is larger. As an example, for a center-fed reflectarray with a broadside directed main beam having an f/D ratio of 0.5, the elements located in the center region of the reflectarray could be designed to have large spacing of 0.95 λ_o, while those elements close to the edge with incident angles around 45° should have spacing less than 0.6 λ_o. However, for ease of design and manufacturing, all element spacing in a reflectarray should be uniform. If all element spacing is 0.95 λ_o, there will be distributed grating lobes forming in the far field.

Distributed grating lobes, rather than a single distinct grating lobe, are formed because the incident angle varies from the center to the edge of the reflectarray. In other words, it would be difficult to physically observe these distributed grating lobes. But they are out there as wasted energy if the element spacing is inappropriately designed. Thus, in order to avoid any distributed grating lobe formation, the element spacing of a reflectarray should be governed by the incident angle of the edge elements. Large f/D ratio generally helps to minimize grating lobe issue. Offset-fed reflectarrays are more prone to grating lobe formation if element spacing is much larger than 0.5 λ_o.

4.2 PATH LENGTH AND PHASE DELAY CALCULATION

The path lengths from the feed to all elements are all different, which leads to different phase delays. To compensate for these phase delays, the elements must have corresponding phase advancements designed in accordance with its uniquely calibrated phase design curve. This curve will be different for different types of elements used. But, they generally have an S-shaped curve, similar to that shown in Fig. 4.2. The following gives an example of how the path length and corresponding required compensating phase are calculated for each element of a reflectarray with a broadside-directed beam. It is basically a simple ray tracing method. The differential path length for each element is given as:

$$\Delta L_{m,n} = L_{m,n} - L_{o,o}, \tag{4.2}$$

where $L_{m,n}$ is the path length between the feed and the mnth element, which can be obtained by a simple geometry calculation. $L_{o,o}$ is the distance between the feed and a reference point on the reflectarray surface, for example, the center point. $\Delta L_{m,n}$ is thus the differential feed path length for the mnth element. To achieve a collimated radiation, the phase advancement $\Delta\Phi_{mn}$ needed for the mnth element is given by

$$\Delta\Phi_{mn} \text{ in degrees} = [\Delta L_{m,n}/\lambda_o - \text{integer of } (\Delta L_{m,n}/\lambda_o)] \times 360. \quad (4.3)$$

The above indicates that the compensating phase can be repeated every 360° and the portion that is integer multiple of a wavelength or 360° can be deleted. These simple equations only demonstrate to the readers how the compensating phases are derived. For other more complicated reflectarray geometries, such as offset-fed or Cassegrain-fed, each designer must have his or her own ray-tracing method to calculate the correct path lengths and, thus, phases.

4.3 RADIATION PATTERN CALCULATION

The far-field radiation patterns of a reflectarray should be accurately calculated so that the main beam width and direction, sidelobe level, cross-polarization level, and directivity can all be accurately predicted for the required system. With all elements' compensating phases known, the far-field radiation patterns can be calculated by the conventional array theory, where the radiations of all elements are summed together as follows. Consider a planar array, shown in Fig. 4.5, consisting of M × N elements that are nonuniformly illuminated by a low-gain feed at position vector $\vec{r}_f$. Let the desired beam direction be specified by unit vector $\hat{u}_o$. Then the far-field of the reflectarray in the $\hat{u}$ direction will be of the form

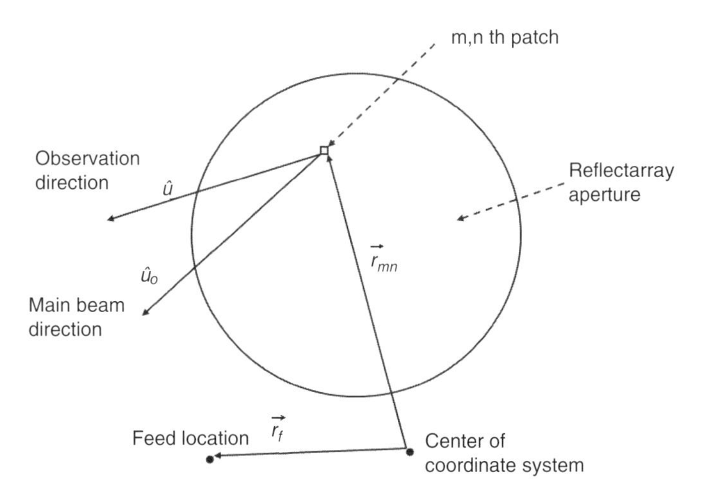

Figure 4.5. Coordinate system of the reflectarray.

$$E(\hat{u}) = \sum_{m=1}^{M}\sum_{n=1}^{N} F(\vec{r}_{mn} \bullet \vec{r}_f)\cdot A(\vec{r}_{mn} \bullet \hat{u}_o)\cdot A(\hat{u} \bullet \hat{u}_o)\cdot \exp[jk(|\vec{r}_{mn} - \vec{r}_f| + \vec{r}_{mn} \bullet \hat{u}) + j\alpha_{mn}] \tag{4.4}$$

where F is the feed pattern function, A is the reflectarray element pattern function, $\vec{r}_{mn}$ is the position vector of the mnth element, and α_{mn} is the required compensating phase of the mnth element, for example, the phase calculated by Equation (4.3) for a broadside beam. A $\text{Cos}^q\theta$ factor is used for both F and A functions with no azimuth (ϕ) dependence. The field pattern calculated by Eq. (4.4) is the reradiated fields from all the elements. Edge diffracted field can be included in this equation by using the Geometrical Theory of Diffraction [9, 10]. Specular reflected field, such as discussed in Section 4.1.4, can also be included to improve the accuracy of the predicted patterns.

The above pattern calculation uses only a conventional array summation technique without including the effect of mutual coupling or mutual scattering between elements. It will yield good main beamwidth, beam direction, and general pattern shape; but will not give accurate sidelobe and cross-polarization predictions. Because a reflectarray generally consists of a huge number of elements, it would be difficult to have a complete rigorous solution to include all mutual coupling effects, even with the modern-day computer capability.

Nevertheless, there are two approximation methods that are quite accurate to calculate radiation patterns. One that has been used quite often is the infinite-array approach [1, 2]. It first calculates the single element pattern of the reflectarray by assuming it is in an infinite array environment with all surrounding elements identical. This single element pattern is then summed together for all the elements by using the formula similar to Equation (4.4) to generate a far-field pattern. Thus, the element function A in Equation (4.4) will contain mutual coupling effects in an infinite array environment, but with some degree of approximation. The second method calculates the single element pattern by including all mutual coupling effects of only the nearby surrounding elements with rigorous Method of Moments solution [11]. The number of nearby elements could range from several to 50, depending on the type of elements used and the main beam angle. These nearby elements are not identical but are the actual exact elements on the reflectarray. The resulting single element pattern is then summed together by using Eq. (4.4) to yield a far-field pattern. This method, although more accurate, is more laborious because all elements with unique surrounding elements throughout the entire array need to be calculated.

4.4 REFLECTARRAY GEOMETRY DESIGN

A good geometry design will yield good reflectarray efficiency. To determine the geometry of either a center-fed or an offset-fed reflectarray is basically governed by its required f/D ratio and its desired aperture efficiency. The aperture efficiency (η_a) can be defined as the product of the illumination (η_I) and spillover (η_s) efficiencies: $\eta_a = \eta_I \times \eta_s$. As an example, by integrating the pattern function of Eq. (4.4), the illumination efficiency for a center-fed reflectarray can be obtained in a closed form [7] as given by

$$\eta_I = \frac{[((1+\cos^{q+1}\theta_e)/(q+1)) + ((1-\cos^q\theta_e)/q)]^2}{2\tan^2\theta_e[(1-\cos^{2q+1}\theta_e)/(2q+1)]}, \tag{4.5}$$

and the spillover efficiency is given by

$$\eta_s = 1 - \cos^{2q+1}\theta_e, \tag{4.6}$$

where q is the exponent of the feed pattern function represented by $\cos^q\theta$ and θ_e is half of the subtend angle from the feed to the reflectarray aperture. The reflectarray element pattern is approximated by a cosine function. Eqs. (4.5) and (4.6) are calculated by assuming a circular aperture only for the demonstration of the design procedures. Similar closed-form equations can be easily obtained for square, rectangular, or elliptical apertures by performing the proper integrations.

To give an example about how Eqs. (4.5) and (4.6) can be utilized to optimize a reflectarray design, Fig. 4.6 shows the calculated curve of spillover and illumination efficiencies versus the feed pattern factor q (feed beamwidth) for a half-meter 32 GHz reflectarray with a fixed f/D ratio of 1.0 (θ_e = 26.6°). It demonstrates that the maximum aperture efficiency is achieved at q = 10.5 or

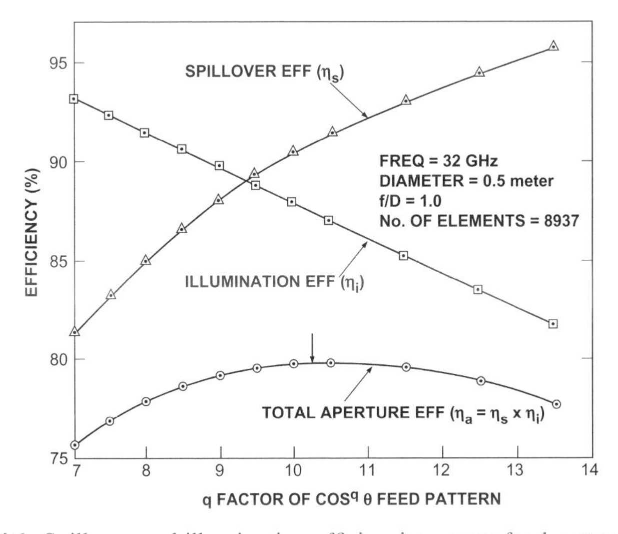

Figure 4.6. Spillover and illumination efficiencies versus feed pattern shape.

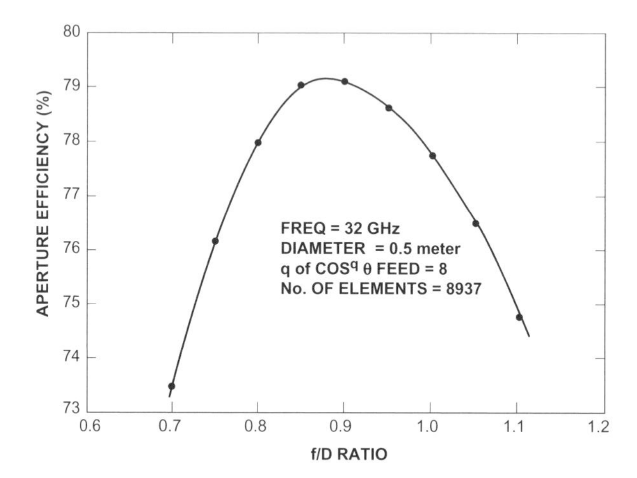

Figure 4.7. Aperture efficiency versus f/D ratio.

when the feed has a –3 dB beamwidth of 29°. Another curve, shown in Fig. 4.7, gives aperture efficiency as a function of f/D ratio for the same half-meter 32 GHz reflectarray when the feed beamwidth is fixed at 33.4° with $q = 8$. In this case, the maximum aperture efficiency is achieved when the f/D ratio is 0.87. It can be seen that curves derived from Equations (4.5) and (4.6) are essential to obtaining an optimum efficiency design. Although the above discussion demonstrated the simple case of a center-fed reflectarray, one should be able to realize the importance of an appropriate geometry design. A more rigorous approach could be used to accurately calculate the efficiencies and thus obtain a truly optimal geometry design.

Another factor of concern when designing the geometry of a reflectarray is the aperture's electrical size. Generally, the smaller the size, the worse the performance in terms of sidelobe and cross-polarization levels, as well as efficiency (assume mechanical issues are not considered here). This is also true for parabolic reflectors; in particular, when the size corresponds to an antenna gain of less than 25 dB, or in the reflectarray case, the number of elements becomes less than a few hundreds. The reason is because, as the aperture size gets larger, the amounts of undesirable edge diffraction, specular reflection, and feed blockage become less relative to the desirable radiation from all the elements. To demonstrate this effect, the radiated co-polarization far-field patterns are calculated using Eq. (4.4) with and without edge-diffracted and specular-reflected fields [8]. The edge-diffracted field is calculated using the Uniform Geometrical Theory of Diffraction (UGTD) [9, 10], while the amount of specular reflected field is determined by using the waveguide simulator approach mentioned in sections 4.1.1 and 4.1.4. Fig. 4.8 presents the calculated

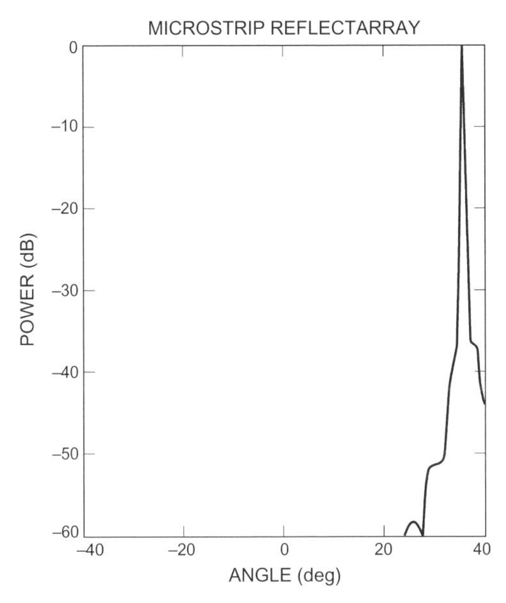

Figure 4.8. Calculated reflectarray pattern without edge diffracted and specular reflected field components, aperture diameter = 144 wavelengths.

pattern for a reflectarray having a diameter of 144 λ_o without including the effects of edge diffraction and reflection. It has the main beam scanned to 36° and shows no far-out sidelobes above −60 dB. With the same antenna geometry, Fig. 4.9 shows the pattern that does include the components of edge diffraction and specular reflection. In this pattern, many sidelobes are near the −40 dB level, which is still considered, in many systems, very low.

Now, let's look at a smaller reflectarray with a diameter of only 10 λ_o. The far-field pattern of this antenna, when the main beam is scanned to 36°, is given in Fig. 4.10 where the edge diffraction and specular reflection are not included. It shows a peak sidelobe level of −27 dB. For the same small reflectarray, the pattern with edge diffraction and specular reflection included is shown in Fig. 4.11 where a peak sidelobe level of −15 dB is observed. Figs. 4.8 through 4.11 have demonstrated that the reflectarray antenna will have good pattern performance when its aperture directivity is significantly higher (e.g., 30 dB) than its feed directivity. In other words, the reflectarray antenna will become more efficient when its aperture is electrically larger. However, on the other hand, as the electrical size of the reflectarray gets large, its bandwidth becomes smaller for the reason explained in Chapter 1, Section 1.2.2.2. The designer must be aware of all these factors when deciding upon the size of the aperture.

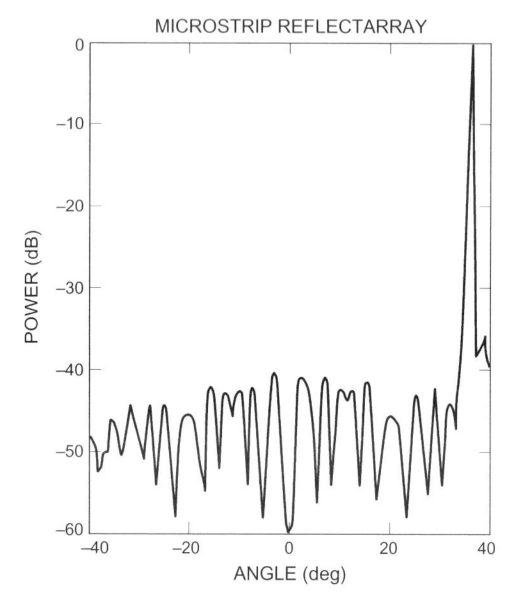

Figure 4.9. Calculated reflectarray pattern with edge diffracted and specular reflected field components, aperture diameter = 144 wavelengths.

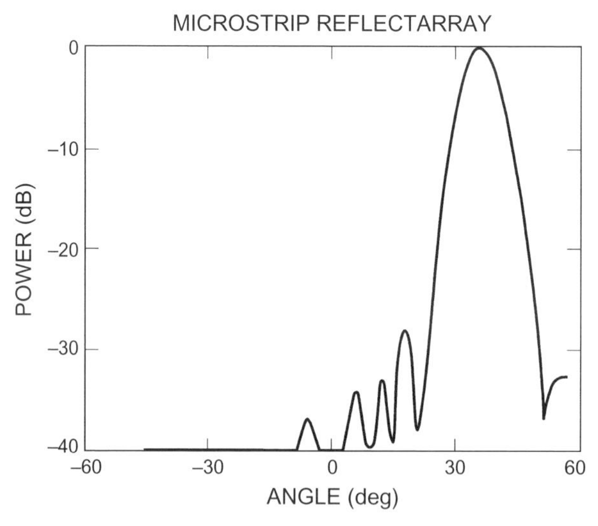

Figure 4.10. Calculated reflectarray pattern without edge diffracted and specular reflected field components, aperture diameter = 10 wavelengths.

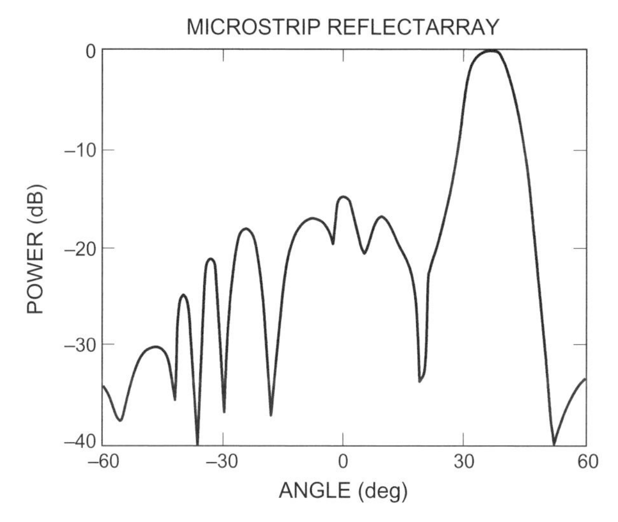

Figure 4.11. Calculated reflectarray pattern with edge diffracted and specular reflected field components, aperture diameter = 10 wavelengths.

4.5 REFLECTARRAY POWER HANDLING

There is a general concern about the power handling capability of a printed reflectarray since the reflectarray elements are often fabricated on thin substrates. On the contrary, a printed reflectarray antenna could generally handle relatively large amount of RF power. This is because, during transmit, the high power from the feed horn will illuminate a large number of reflectarray elements and, thus, the per-element transmitted power is low. As an example, a 3-m X-band reflectarrray operating at 10 GHz would have about 31,000 elements. A one Kilowatt transmit power would yield a per-element power of about 32 milliwatt, which should not be a problem for printed microstrip elements with nominal substrate thickness.

REFERENCES

1. D. M. Pozar and T. A. Metzler, "Analysis of a reflectarray antenna using microstrip patches of variable size," *Electronics Letters*, April 1993, pp. 657–658.
2. D. Pozar, S. D. Targonski, and H. D. Syrigos, "Design of millimeter wave microstrip reflectarrays," *IEEE Trans. Antennas Propagat.*, Vol. 45, February 1997, pp. 287–296.
3. E. Girard, R. Moulinet, R. Gillard, and H. Legay, "An FDTD optimization of a circularly polarized reflectarray unit cell," IEEE AP-S/URSI symposium, San Antonio, Texas, June 2002, pp. III-136–139.

4. J. A. Encinar, "Design of two-layer printed reflectarray using patches of variable size," *IEEE Trans. Antennas Propagat.*, Vol. 49, October 2001, pp. 1403–1410.
5. J. A. Encinar and J. A. Zornoza, "Three-layer printed reflectarrays for contour beam space applications," *IEEE Trans. Antennas Propagat.*, Vol. 52, May 2004, pp. 1138–1148.
6. D. C. Chang and M. C. Huang, "Multiple polarization microstrip reflectarray antenna with high efficiency and low cross-polarization," *IEEE Trans. Antennas Propagat.*, Vol. 43, August 1995, pp. 829–834.
7. J. Huang, "Analysis of a microstrip reflectarray antenna for microspacecraft applications," JPL TDA Progress Report No. 42-120, February 15, 1995.
8. J. Huang, "Microstrip reflectarray antenna for the SCANSCAT radar application," JPL Publication No. 90-45, November 15, 1990.
9. R. G. Kouyoumjian and P. H. Pathak, "A uniform geometrical theory of diffraction for an edge in a perfectly conducting surface," Proceedings IEEE, Vol. 62, Nov. 1974, pp. 1448–1461.
10. J. Huang, "The finite ground plane effect on the microstrip antenna radiation patterns," *IEEE Trans. Antennas Propagat.*, Vol. AP-31, July 1983.
11. M. A. Milon, R. Gillard, D. Cadoret, and H. Legay, "Comparison between the infinite-array approach and the surrounded-element approach for the simulation of reflectarray antennas," IEEE AP-S/URSI symposium, Albuquerque, New Mexico, July 2006.

CHAPTER 5

Broadband Techniques

As mentioned in Chapter 1, the main limitation to reflectarray performance is the narrow bandwidth, generally lower than 5 percent and even less for large reflectarrays. Bandwidth limitation is an inherent characteristic of reflectarrays and much effort has been made in recent years to improve the bandwidth (see the references in this chapter). Reflectarray bandwidth is mainly limited by two different factors as described in [1]. The first one is the narrow band of the radiating elements, and the second one is the differential spatial phase delay resulting from the different lengths from the feed to each point on the wave front of the radiated beam. Both effects are discussed in this chapter and some possible techniques for bandwidth improvement are addressed.

5.1 BANDWIDTH LIMITATION BY THE REFLECTARRAY ELEMENT

The bandwidth limitation produced by the radiating element is the most stringent for moderate size reflectarrays, as demonstrated in [2]. Microstrip patches on a thin dielectric layer are usually used as radiating elements, and their narrow band (3–5 percent) limits the bandwidth of the reflectarray. However, the bandwidth of the radiating element in a printed array is not exactly the same as in a reflectarray. The bandwidth in a printed array is normally determined by the matching of the input impedance of the element, and it is defined by a maximum value of return losses.

Good impedance matching at the radiating element-stub interface is important for reflectarray elements with attached or aperture-coupled stubs, in order to ensure that most of the incident energy is coupled to the stubs and reradiated with the appropriate phase-shift. In principle, a reflectarray element that is well matched to the phasing stub in a given frequency band will provide the appropriate phasing in the whole band. However, the phase response is distorted as a result of the resonances of the combined structure produced by the patch and line. For this reason, the phase curves versus stub length (including both the reradiated and reflected fields) are not linear as expected from

Reflectarray Antennas, by John Huang and José A. Encinar

transmission line theory. See Fig. 3.6, which has been reproduced here as Fig. 5.1 for clarity.

The same phenomenon has been observed in aperture-coupled patches. When the geometrical dimensions of the aperture, slot, and matching stub are adjusted for a good match from the microstrip line, the phase curves exhibit a nonlinear behavior produced by the resonances, as already discussed in Chapter 3 and illustrated in Figs. 3.20 and 3.21. In conclusion, a good match of the radiating element is desirable for good phase control using varying length stubs, but does not ensure a good reflectarray bandwidth. Note also that the input impedance bandwidth of the elements makes no sense when varying-sized patches are used.

When the phase-shift is obtained by varying the resonant length of printed patches, a thin dielectric (around one tenth of a wavelength) should be used in order to achieve a sufficient range of phase-shift (>300°). As a result of the narrow band of the radiating element, the reflection-phase curves versus the patch length are highly nonlinear, showing a steep slope near resonance and very slow variations near the extremes. The phase curves for an array of square patches are shown in Fig. 5.2 at three frequencies. The reflection phase is very sensitive to frequency variations near resonance and almost independent near the extremes as a result of the S-shaped curve. Consequently, the phase distribution on a reflectarray with patches of different sizes will change with frequency and the radiation patterns will be distorted, limiting the operational frequency band of the reflectarray.

In a first approach, an ideal broadband reflectarray element should provide phase curves (versus length) at different frequencies parallel to each other and independent of the angle of incidence. Again, these phase curves should include both the field reradiated by the patches and the one reflected by the

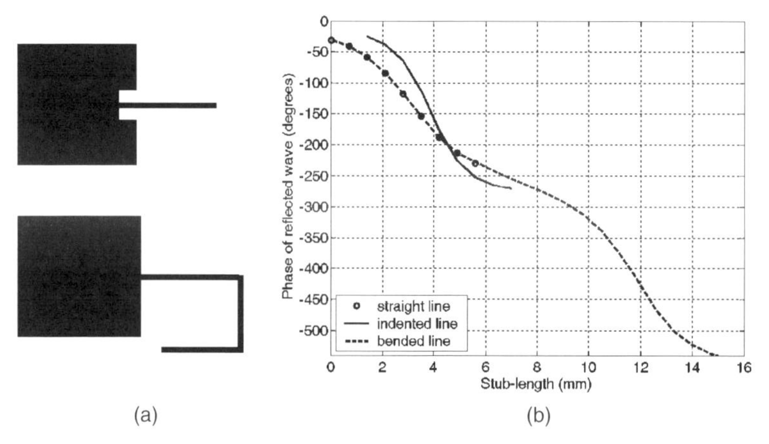

Figure 5.1. Phase response for patches with attached stubs. (a) Element. (b) phase for normal incidence versus the stub-length (t = 1.59 mm, ε_r = 3.2).

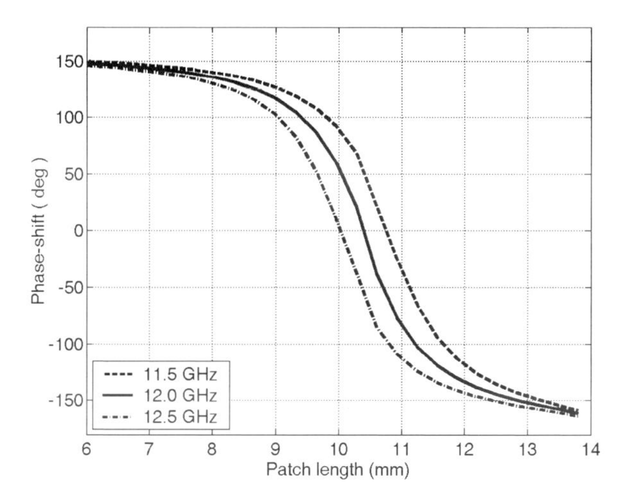

Figure 5.2. Phase-shift at normal incidence for a periodic array of square patches on a grounded substrate versus the patch side a_1 at three frequencies ($p_x = p_y = 14\,\text{mm}$, $t = 1\,\text{mm}$, $\varepsilon_r = 1.05$).

grounded substrate. For these ideal elements, the phase-shift distribution introduced by the reflectarray will only vary by a constant when the frequency changes. However, in this case, the phase of the reflected field will vary in a slightly different manner when changing the frequency, because the phase of the incident wave changes as a result of the different distances from the feed to the element. This effect is discussed in Section 5.3 and illustrated in Fig. 5.6. Then, in a more accurate approach, an ideal reflectarray element should produce a phase-shift that varies with the frequency in the same manner as the phase delay produced by a parabolic reflector (Fig. 5.6). The phase-shift for this ideal element should be proportional to the frequency and to the length of the patch or line (depending of the element type), as in an open-ended transmission line ($-2\beta_l L$, where β_l is the propagation constant in the line). According to previous considerations, a good bandwidth of a reflectarray element is obtained when the phase difference at extreme frequencies varies smoothly. It should be almost constant or preferably slightly increasing with phase magnitude. The bandwidth of the radiating element can be improved by an appropriate design of the phase-shifter element as discussed in the next section.

5.2 BROADBAND PHASE-SHIFTER ELEMENTS

Two types of reflectarray elements that have demonstrated their ability to achieve smooth phase curves, and thus improve the element bandwidth, are studied in this section. The first is a phase-shifter element based on patches with aperture-coupled stubs [3–7], which can be designed to provide a linear variation of phase-shift versus the stub length within a given frequency band. The second reflectarray element consists of two-stacked varying-sized patches and allows smooth phase curves in a range larger than 360° [8–10]. Both reflectarray elements provide a bandwidth larger than 10 percent. Finally, other types of reflectarray elements proposed for bandwidth improvement are reviewed.

5.2.1 Aperture-Coupled Patches

The reflectarray element with the phasing stub aperture-coupled to a radiating printed patch is shown in Fig. 5.3(a). Its operating principle, design, and analysis were addressed in Chapter 3. However, it has already been mentioned that the design of the radiating element for very low return losses seen from the delay line is not sufficient to ensure good bandwidth behavior [4–6].

For a broadband behavior of the reflectarray element, the phase curves versus stub length should be smooth and almost parallel at different frequencies. In the aperture-coupled element there are more geometrical parameters that can be adjusted as additional degrees of freedom to improve the linearity of the phase curves, as the length of the aperture or the dimensions of the patch. The length of the aperture was adjusted in Section 3.7.2 to compensate for the nonlinear behavior of the phase response [7]. Starting from the geometrical parameters obtained for a matched aperture-coupled element, the matching stub was tuned slightly and the slot length varied from 5.57 mm to

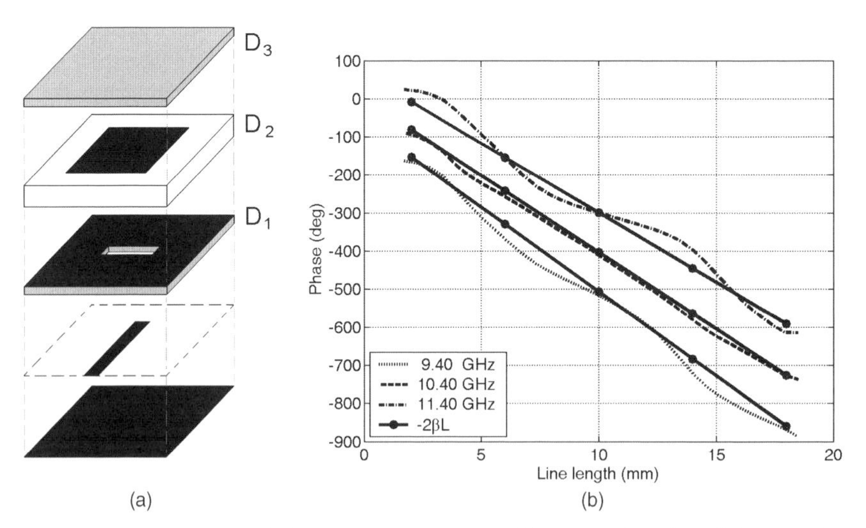

Figure 5.3. Aperture-coupled reflectarray element. (a) Expanded view, (b) phase delay versus line length.

9.57 mm. For the geometrical data given in Table 3.2 of Chapter 3, the slot length was chosen as 6.77 mm, because it provides a very linear phase curve for the delay line varying from 0.00 mm to 18.49 mm. The phase curve for normal incidence obtained by the full-wave method described in Chapter 3 is reproduced here at three frequencies. At the central frequency (10.4 GHz), the phase practically coincides with the ideal phase-delay ($-2\beta L$) in two 360° cycles. The curves at the extreme frequencies in a 10 percent bandwidth were shown in Fig. 3.29 in Chapter 3, which showed a behavior very close to that of the ideal element. For a 20 percent bandwidth the phase curves are still smooth and close to the ideal phase-shifter element as shown in Fig. 5.3.

5.2.2 Variable-Sized Stacked Patches

The phase curves of a single layer of varying-sized patches limit the reflectarray bandwidth because of the nonlinear S-shape shown in Fig. 5.2. A smoother phase variation can be obtained by increasing the dielectric thickness, but in that case the total phase range is reduced to less than 300° and the element is not valid for designing a reflectarray, in which all phase values are required.

The objective of achieving a smooth phase variation within a range larger than 360° can be obtained by stacking two or more arrays, as explained in the following. An array of metallic patches behaves as a resonant circuit, in which the phase of the reflected wave varies when the frequency or the resonant length changes. The maximum phase variation in a resonant element (in front of a ground plane) is theoretically limited to the range ±180°. In the reflectarray element, the phase range is reduced by the phase delay produced by the wave traveling back and forth in the grounded substrate when the patch dimensions are very small, or when the patch is removed. Then, the total attainable range in phase-shift for an array on a dielectric of thickness t is approximately $(2\pi - 2K_\varepsilon t)$ in radians, where K_ε is the propagation constant in the dielectric. Therefore, the range is reduced for thicker substrates, which on the other hand are required for a smoother phase curve. For example, the phase range is reduced to 180° if the dielectric thickness is a quarter-a-wavelength in the dielectric. By stacking several arrays, a multiresonant behavior is obtained and the theoretical limit for the phase range becomes several times 360°. Then, the dielectric thickness of each layer can be increased to obtain a smoother and more linear phase variation, as demonstrated in [8–10].

To improve the bandwidth of the two-layer configuration shown in Fig. 5.4(a), the thickness of separators t_1 and t_2 and the relative size of the patches in each layer are adjusted in order to achieve a sufficiently linear phase variation as a function of the patch dimensions for different incidence angles, for different frequencies, and in a range greater than 360°. Fig. 5.4(b) shows the phase curves at normal incidence for a two-layer element, where the period is 14 mm × 14 mm and the separators are the 3-mm thick Rohacell HF-31. Note that the phase curves are smooth in the 11–13 GHz frequency band (16 percent). The curves at different frequencies are almost parallel, but with a

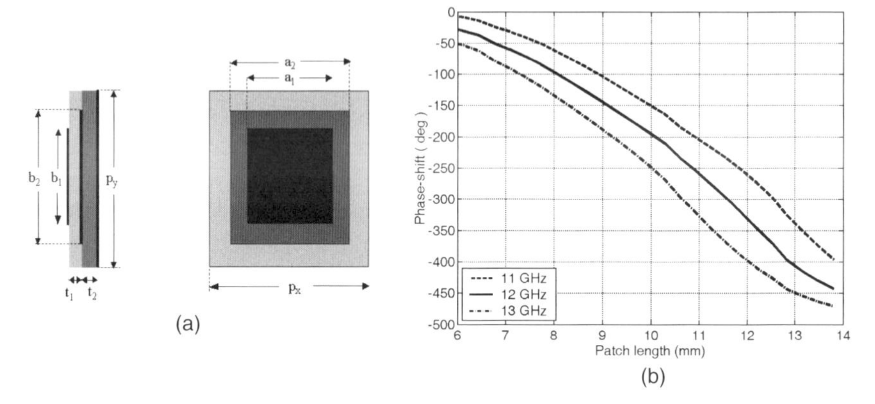

Figure 5.4. Two-layer reflectarray element with varying sized patches. (a) Periodic cell, (b) phase-shift versus patch side ($a_1 = b_1$, $a_2 = b_2$, $p_x = p_y$ =14 mm, $t_1 = t_2 = 3$ mm, $\varepsilon_r = 1.05$, $a_1 = 0.7a_2$).

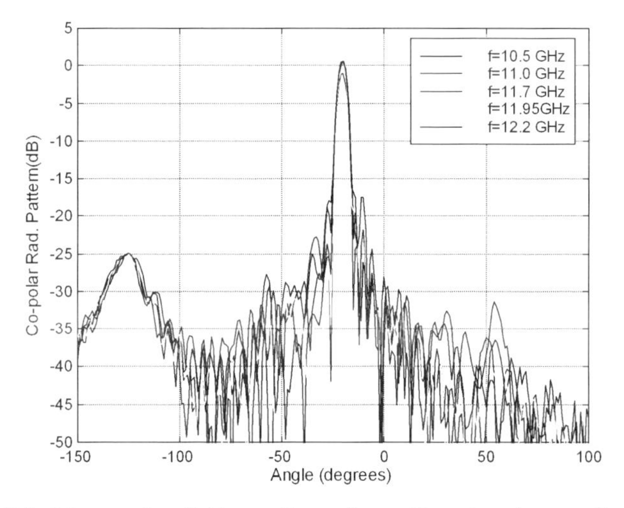

Figure 5.5. Measured radiation patterns for a 40-cm two-layer reflectarray.

slight diverging behavior when the phase-shift increases, as it should be in an ideal phase-delay element, where the phase is $-2\beta L$.

Based on the previous two-layer element, a 40-cm reflectarray was designed, built, and measured. The geometrical parameters and a photo of the breadboard are given in Section 3.8.3 of Chapter 3 and in more detail in reference [9]. The measured radiation patterns are shown in Fig. 5.5 at different frequencies to check the bandwidth for E_x polarization (with the electric field in the

TABLE 5.1. Comparison of two-layer and one-layer reflectarrays

	Reflectarray in [11]	Reflectarray in [12]	Two-layer prototype	
Central freq. (GHz)	9.75	9.075	11.95	
Frequency band (GHz)	9.5–10.0	8.85–9.5	11.5–12.4	11–13
Frequency band (%)	5.1	7.2	7.5	16.7
Polarization	linear	dual linear	dual linear	
F/D	0.9	0.87	0.84	
Radiation angle (degrees)	25	30	19	
Surface/λ^2	261.6	383.0	205.5	
Maximum directivity (dBi)	34.7	36.2	33.8	
Measured directivity (dBi)	—	—	32.47	
Measured gain (dBi)	30.5	33	31	
Gain variations (dB)	1.5	3	0.3	1.5
Cross-polar level (dB)	≤–17	≤–25	≤–25	

reflectarray x-direction). At the extreme frequencies, the radiation patterns are similar, with some increase in sidelobe levels and small gain variations. Note that the sidelobe of –25 dB at –125° is the direct radiation from the feed (spillover). A 16.7 percent bandwidth (11–13 GHz) is achieved within 1.5 dB gain variations, while a very stable gain was measured (31 ± 0.15 dB) within the 11.5–12.4 GHz band. The main performance of the prototype is compared in Table 5.1 with the measured results of other X-band one-layer reflectarrays described in the literature [11], [12].

The phase shifter elements used in the breadboards are microstrip dipoles in [11] and patches with attached stubs in [12]. These results show that the two-layer configuration provides a larger bandwidth than the other single-layer reflectarray concepts. To compare the gain performance, it must be taken into account that the electrical surface is smaller for the two-layer prototype. For an identical electrical size, the two-layer reflectarray will provide a larger gain and efficiency, since the difference between measured gain and maximum directivity (given by the electrical surface) is smaller than in the other references.

5.2.3 Other Reflectarray Elements for Bandwidth Improvement

Different types of reflectarray elements have been proposed in recent years to improve the element bandwidth in printed reflectarrays [13–21]. Several geometries of printed ridged-shaped patches with varying dimensions were analyzed in [13], and the phase-shift performances were compared with those for classic elements (rectangles, dipoles, and rectangles with tuning stubs). The ridge-type patches have a slightly better performance in phase-shift than a single layer of rectangular patches, but the irregular shape will increase the cross-polarization.

Double-layer structures were also analyzed in [13], showing a superior performance with respect to single-layer configurations. Stacked metallic rings were proposed as a reflectarray element in [14]. As in rectangular stacked patches, the reflection phase is controlled by varying the size of the printed rings. The resonant frequencies of the two arrays are adjusted to reduce the gradient and improve the linearity of the reflection phase versus ring size [15]. Bandwidth is improved for the stacked ring configuration, but the results are not superior to those shown in Fig. 5.4 for the stacked squared patches. A new reflectarray element made up of double-crossed loops of variable lengths printed on a conductor-backed substrate has been demonstrated in [16]. A single-layer reflectarray was manufactured and measured, which showed a radiation efficiency close to 55 percent and a 1 dB gain–bandwidth of 10 percent, centered at 22 GHz. This bandwidth is larger than in conventional single-layer reflectarrays, but still smaller than the bandwidth obtained for a two-layer reflectarray [9].

A reflectarray element based on proximity-coupled stubs was proposed in [17], which allows a certain bandwidth improvement. However, the bandwidth improvement was mainly produced by the use of a thicker substrate as clarified in [18]. Electromagnetic coupling was also used in [19] for a QUAD-EMC element, which consists of four identical microstrip patches coupled electromagnetically by a driven patch on a lower substrate. The advantage of this configuration is that only one phase control is used for four radiating patches. Experimental results showed that this concept is valid and the performance can be similar to those of proximity coupled stubs [17]. A 23 percent bandwidth was reported for a breadboard, however, the efficiency was very low (21 dB of gain for 27 dBi of maximum directivity). The difference between measured gain and maximum directivity is 6 dB, which is extremely high when compared with the 2.8 dB difference obtained in the breadboard reported in [9]. Therefore, the large bandwidth can be attributed to a significant reduction in gain at all frequencies.

Rectangular patches loaded with slots have been proposed as a reflectarray element, where the reflection phase is controlled by adjusting both, the length of the patch, and the length of the slots [20–21]. The range of phase-shift was increased up to 550° using two coupled patches with slots as demonstrated in [20]. This type of element was used to design, manufacture, and test a reflectarray antenna in Ku-band [21] with 437 elements, but the measured bandwidth was only 4 percent for a focused beam.

5.3 BANDWIDTH LIMITATION BY DIFFERENTIAL SPATIAL PHASE DELAY

Bandwidth in large reflectarrays, as in the case of space applications, is drastically reduced because of the different path lengths, as discussed in Chapter 1. Parabolic reflector antennas use the physical curvature to equalize signals arriving from the feed and to form a planar wave front. This is the true time delay compensating method, which in principle is frequency independent. In contrast, printed elements of the reflectarray only allow for phase-shift compensation (in a range of 360°). For large apertures, the path length that produces the phase delay in a parabolic reflector can be on the order of many wavelengths. However, the phase-shift in the reflectarray elements is limited to a 360° range, and the phase delay is only exactly compensated at the central frequency. The phase error at a different frequency increases for large values of phase delay, and the phase compensation is valid only over a very narrow frequency band. This band limitation is the same that occurs in zoned reflectors [22], in which the depth of a parabolic reflector is reduced in steps of an integer number of half-wavelengths, in a similar manner as in zoned dielectric lenses [23].

To evaluate the effect of compensating phase-shift instead of phase delay, some phase definitions are made considering a flat reflectarray superimposed on top of an "equivalent parabolic reflector" with its focus on the phase center of the feed (Fig. 5.6). The phase delay required at element i of the reflectarray to generate a focused beam in the direction of the paraboloid axis will be

$$\phi_{di}(f) = -K_0(d_{1i} + d_{2i} + d_0), \tag{5.1}$$

where the FAB triangle in Fig. 5.6 is isosceles, K_0 is the propagation constant in vacuum, and d_0 is a constant to fix the phase delay on the reflectarray contour to a realizable value. For example, d_0 can be chosen so that $-K_0d_0$ is the minimum phase-shift obtainable with the reflectarray element, that is, with

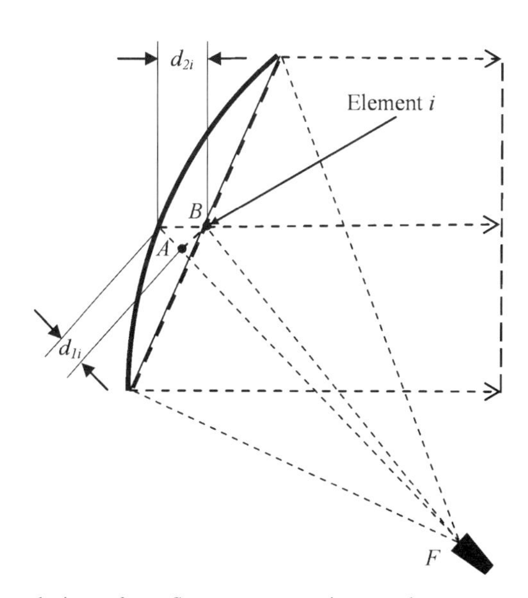

Figure 5.6. Lateral view of a reflectarray superimposed onto a parabolic reflector.

a zero-length stub or with a small patch in varying-sized elements. Eq. (5.1) gives the phase delay produced by the equivalent parabolic reflector on the reflectarray plane. The same phase delay distribution can be obtained in the following alternative manner. To produce the same focused beam as that of the parabolic antenna, in the direction (θ_b, φ_b) using the coordinate system centered on the reflectarray as shown in Fig. 3.1, the progressive phase distribution given by Eq. (3.1) is needed on the reflectarray surface. Then, the required phase delay on the reflectarray plane is obtained by subtracting from the previous distribution, the delay that the wave experiences when traveling from the focus to element i ($-K_0 d_{fi}$), being expressed as

$$\phi_{di}(f) = -K_0 \sin\theta_b (\cos\varphi_b x_i + \sin\varphi_b y_i) + K_0 d_{fi} - K_0 C, \tag{5.2}$$

where, (x_i, y_i) are the coordinates of the element i and C is a constant to fix the phase delay on the reflectarray contour to a realizable value. The phase delay for a 1.5-m offset antenna with an f/D ratio equal to one and 200 mm offset distance (from the paraboloid axis to the reflectarray edge) is shown in Fig. 5.7, which varies in a range from 0 to −2438°. After limiting the phase to a 360° range, the phase-shift given in Fig. 5.8 is obtained.

For an ideal parabolic reflector, the phase delay on the reflectarray plane will be directly proportional to the frequency, as shown in Eqs. (5.1) and (5.2). Then, the difference in phase delay at frequencies f_l and f_0 for the element i should be

$$D_{di}(f_l, f_0) = \phi_{di}(f_l) - \phi_{di}(f_0) = \frac{f_l - f_0}{f_0} \phi_{di}(f_0). \tag{5.3}$$

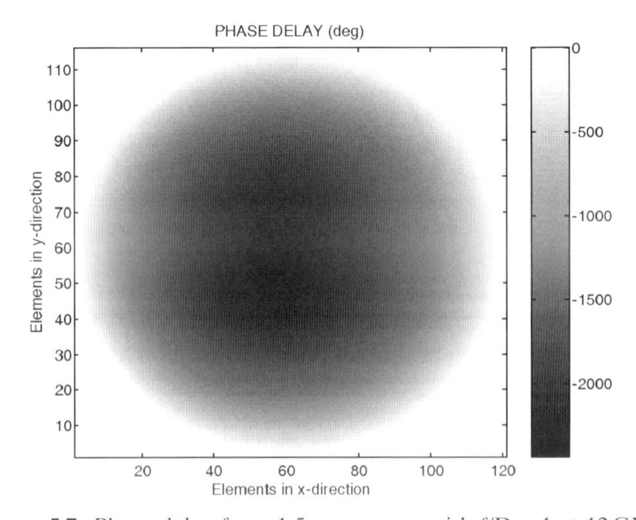

Figure 5.7. Phase delay for a 1.5-m antenna with f/D = 1 at 12 GHz.

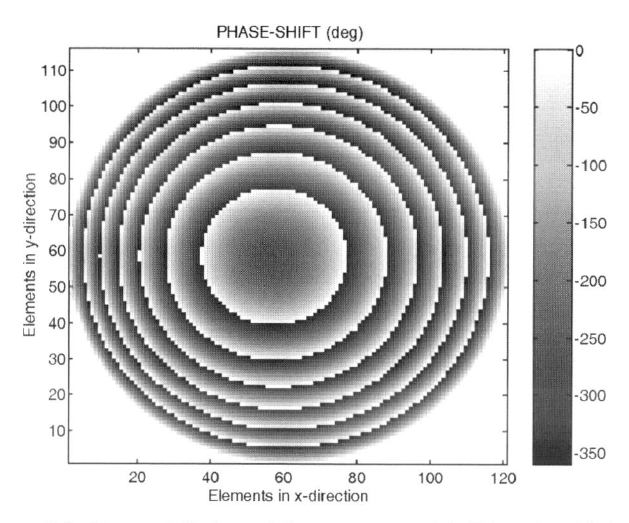

Figure 5.8. Phase-shift for a 1.5-m antenna with f/D = 1 at 12 GHz.

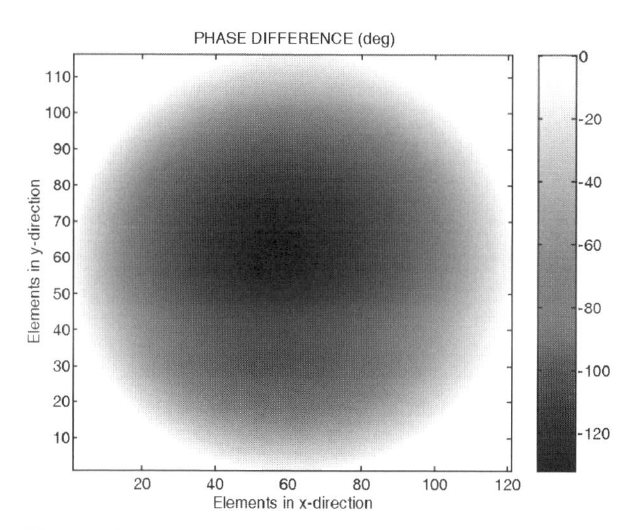

Figure 5.9. Difference in phase delay for a 1.5-m antenna with f/D = 1 at 12.6 GHz and 12.0 GHz.

The difference in phase delay at 12.6 GHz and 12.0 GHz for the 1.5-m antenna is shown in Fig. 5.9. Let us consider that an ideal reflectarray element is used to produce the required phase-shift at the central frequency, and that the phase-shift is the same at any frequency. Then, the error in phase at a frequency f_l will be the required difference in phase delay directly as given in Eq. (5.3), which is shown in Fig. 5.9 at 12.6 GHz for the 1.5-m antenna. The radiation patterns are computed at extreme frequencies in a 10 percent bandwidth, assuming the

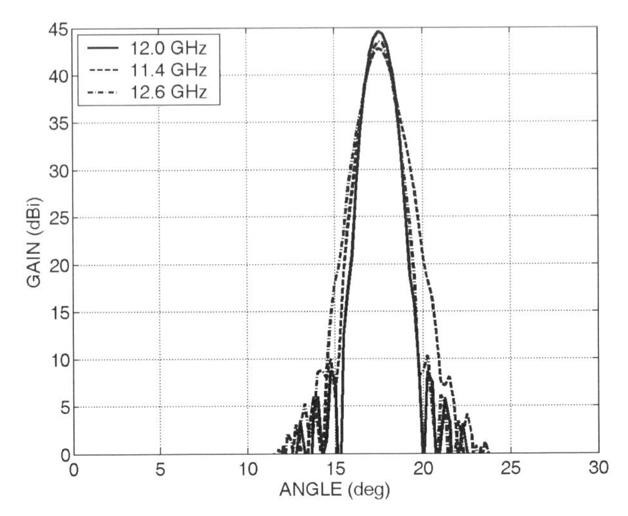

Figure 5.10. Radiation pattern in the offset plane for a 1.5-m reflectarray with f/D = 1 assuming ideal phasing elements.

same phase-shift for all the frequencies and a taper of −18 dB at the edges of the reflectarray. A reduction in gain of 2 dB is obtained at extreme frequencies, compared with the case of a parabolic reflector (Fig. 5.10). Note that the beam is directed at the 17.5° angle using the reflectarray coordinate system.

The radiation patterns are practically unchanged when the phase shift behaves as the ideal broadband element defined in Section 5.1, where the phase-shift varies proportionally to the frequency ($-2\beta_l L$) in the 360° range, because the bandwidth limitation is produced by the phase compensation in 360° zones. The effect of the spatial phase delay is more significant for larger reflectarrays, for smaller f/D ratios, and also for a center-fed configuration, because the distance from the reflectarray plane to the equivalent parabolic surface increases. Fig. 5.11 shows the radiation patterns for a 1.5-m centered-fed reflectarray with an f/D equal to 0.7. In this case, the effect of the spatial phase delay is a reduction of 3 dB at the extremes of a 10 percent bandwidth. These results demonstrate that, for a focused beam, the effect of the differential phase delay is mainly a small reduction in gain. However, for contoured beam reflectarrays, this effect will produce a more severe distortion of the beam shaping, as will be discussed in Chapter 7.

5.4 BROADBAND TECHNIQUES FOR LARGE REFLECTARRAYS

Previous results show that the effect of different path lengths is only significant for very large electrical dimensions, as in the case of antennas for space applications. In the case of very large antennas, several techniques are discussed in this section for bandwidth improvement.

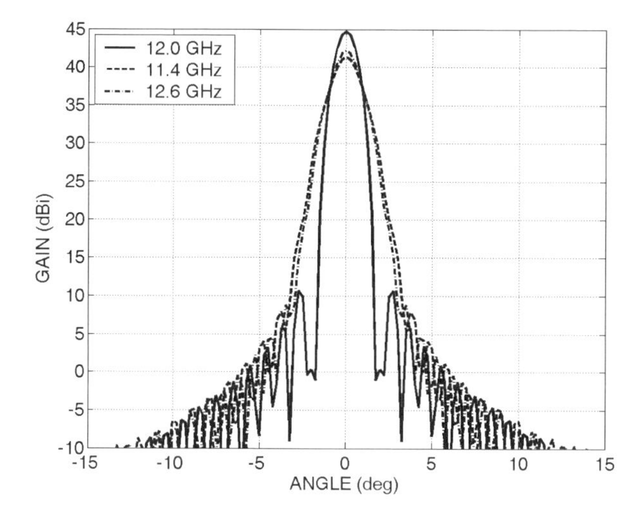

Figure 5.11. Radiation pattern for a 1.5-m centered-fed reflectarray with f/D = 0.7 assuming ideal phasing elements.

The first one is to implement delay lines in order to compensate the real phase delay in the whole range (several times 360°). This approach can be implemented using stubs with a length that varies in a range of several wavelengths as demonstrated in [5]. The inconvenience of this implementation is that more room is required for the long delay lines.

The second technique consists of compensating the spatial phase delay in a given frequency band with the phase of the reflection coefficient. The technique was demonstrated in [10] using a three-layer printed reflectarray with rectangular patches of variable size. The three stacked patches give more degrees of freedom and allow one to optimize the patch dimensions in order to match, not only the phase-shift at the central frequency, but also its variation with frequency. Using this technique, a 10 percent bandwidth was achieved for a 1-m reflectarray at 12 GHz with 0.5 dB gain variations, by compensating the phase delay in a range of five times 360° [10].

Previous techniques can be used to compensate the phase delay for certain antenna dimensions, defined by a limited number of 360° cycles. The limit for the delay lines is imposed by the room available for the stub length, while for the three-layer reflectarray the limit is given by the possibility of compensating the phase delay with the three-stacked patches. Previous results showed that the compensation of phase delay in a 10 percent bandwidth can be achieved for reflectarray apertures up to 70 wavelengths, for an f/D of around one.

For larger reflectarrays, a possible solution to achieve a bandwidth around 10 percent consists of a faceted configuration that approximates the shape of a parabolic surface so that the number of 360° cycles is limited in each flat facet [27], where each planar facet is a reflectarray that introduces the appropriate phase-shift to simulate a real parabolic reflector.

5.4.1 True Time Delay Reflectarrays

The operation bandwidth in large reflectarrays is reduced by the fact of limiting phase compensation to a single 360° cycle, as explained in previous section. In theory, this effect will be eliminated if the reflectarray elements are capable of introducing a real phase delay, as that given by Eq. (5.1) in a range of several 360° cycles. This is called true time delay compensation by analogy with phased-array antennas. Rectangular patches with aperture-coupled lines can produce a true-time delay, where the phase of the reflected wave is directly proportional to the length of the line, if they are appropriately designed as shown in Section 5.2.1. The total range of phase delay that can be achieved with this type of element is only limited by the room available for the lines. The range in phase delay can be enlarged by using a high dielectric constant for the line substrate, and also by bending the line to increase the physical length [5, 6]. Phase delay was achieved in a range of up to 1200° (three 360° cycles) using a substrate with $\varepsilon_r = 10$ [7]. The geometrical parameters of the reflectarray element are given in Table 5.2 and the phase delay at 10.4 GHz is shown in Fig. 5.12 for a periodic cell of $18 \times 18\,\text{mm}^2$. The reflection losses in dB are also plotted in Fig. 5.12(b) with and without an optional ground plane separated by a quarter wavelength to eliminate the leakage of energy transmitted through the apertures (Fig. 5.3(a)). When the ground plane is present, the reflection losses are mainly produced by the dissipation in the dielectric layers, and they are usually less than 0.5 dB in most of the phase range.

The other alternative to increase the range of phase delay is to bend the line [6]. Then, the phase curves shown in Fig. 5.3(b) for a substrate with $\varepsilon_r = 3.2$ can be extended by using a U-shaped delay line, as shown in Fig. 5.13. The range of phase delay is increased from 640° to 1800° (five 360° cycles). The dissipative losses increase slightly with the line length, but they are still less than 0.5 dB. By combining the use of U-shaped delay line and a high dielectric substrate, the phase delay can be adjusted in a range larger than seven 360° cycles, at the expense of a slightly higher loss. Then, the phase delay

TABLE 5.2. Geometrical definition of aperture-coupled reflectarray element

Layer	ε_r	$\tan\delta$	Thickness [mm]	X (mm)	Y (mm)
Line	—	—	0.000	$W_l = 1.5$	$L_s = 0.5$
D1	10.0	0.0030	0.508	—	—
Slot	—	—	0.000	5.37	1.0
D2	1.067	0.0002	2.000	—	—
Patch	—	—	0.000	9.0	9.0
D3	3.2	0.0030	0.508	—	—

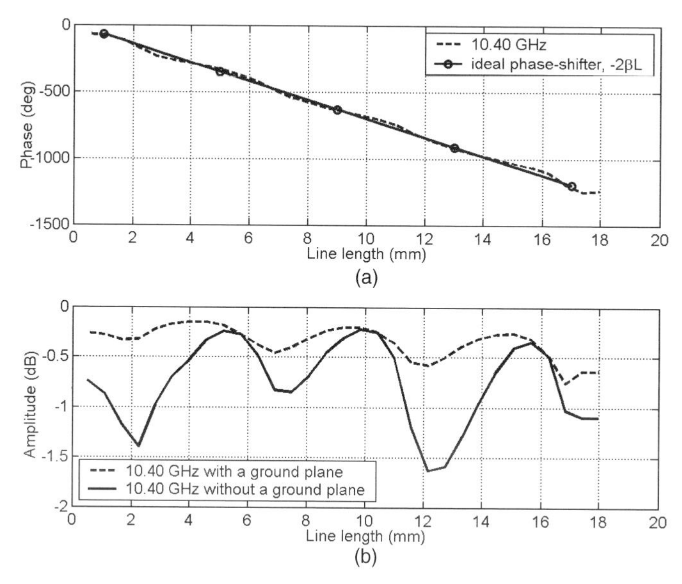

Figure 5.12. Reflection coefficient versus line length. (a) Simulated phase delay compared with $-2\beta L$. (b) Reflection losses with and without a ground plane.

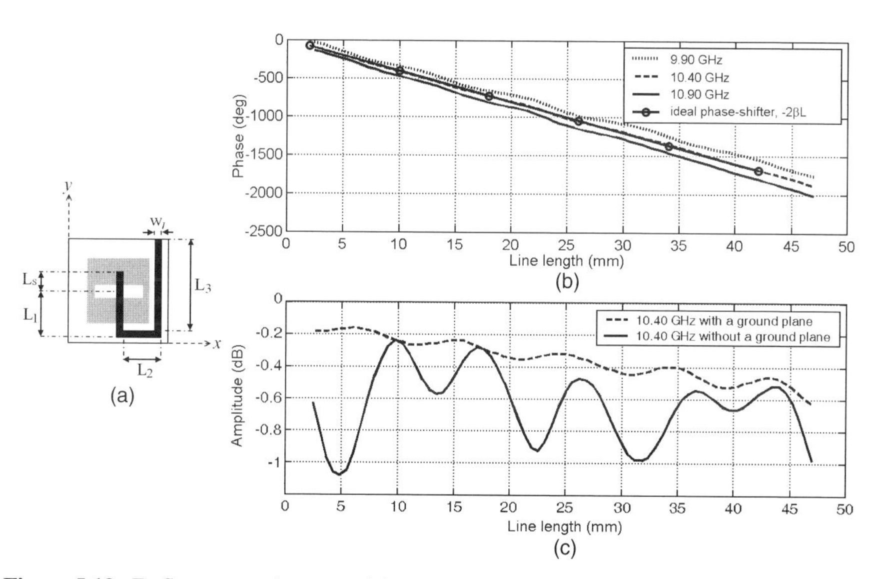

Figure 5.13. Reflectarray element with a U-shaped delay line. (a) Top view of the unit cell. (b) Simulated phase delay compared with $-2\beta L$. (c) Reflection losses with and without a ground plane.

shown in Fig. 5.7 for a 1.5-m reflectarray at 12 GHz can be implemented using true time delay compensation. In this case, the bandwidth of the reflectarray will only be limited by the previously discussed element phasing performance. The aperture-coupled reflectarray elements have been designed for linear polarization, but the same element can be used for dual linear, or circular, polarization by adding additional apertures and delay lines for the orthogonal polarization, as shown in Fig. 5.14.

5.4.2 Compensation of Phase Delay in a Frequency Band

Another alternative to overcome the bandwidth limitation produced by the different patch lengths consists of compensating the phase delay given by Eq. (5.1) with the phase of the reflection coefficient on each reflectarray element at different frequencies. That means that the reflectarray elements should be optimized to provide not only the required phase-shift at central frequency, but also an appropriate variation with frequency to compensate the phase delay at other frequencies [10].

The phase delay $\phi_{di}(f)$ required at element i in a flat reflectarray to generate a focused beam (Eq. 5.1) was derived to compensate twice the path distance between a parabolic and a flat surface (Fig. 5.6). The phase delay, which is not limited to a 360° range, was used to define the phase differences at two different frequencies. The difference in phase delay at extreme frequencies (11.4 GHz and 12.6 GHz) for the 1.5-m antenna with f/D = 1 is shown in Fig. 5.15. To compensate the phase delay in a given frequency band, the reflection phase of the reflectarray elements should match the required phase-shift (limited to 360° range) at central frequency ϕ_{oi} (Fig. 5.8) as well as the difference in phase delay at extreme frequencies $D_{di}(f_1, f_2)$ (Fig. 5.15). The phase of the reflection

Figure 5.14. Aperture-coupled reflectarray elements for dual polarization.

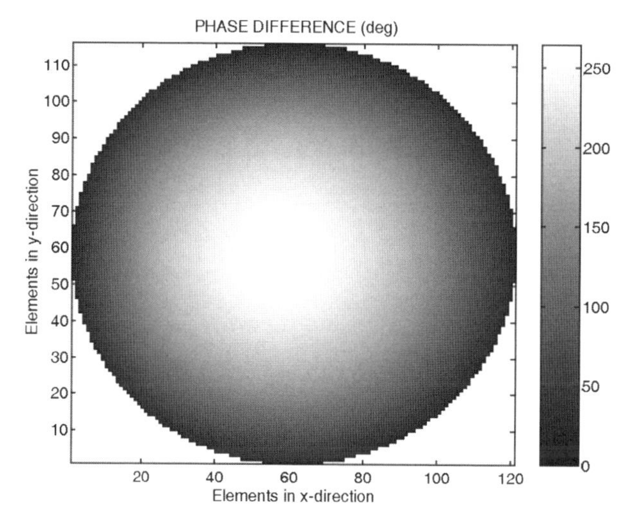

Figure 5.15. Difference in phase delay for a 1.5-m antenna with f/D = 1 at 11.4 GHz and 12.6 GHz.

coefficient, $\phi_r(f)$, and its variations with frequency are analyzed for different multilayer configurations with two and three stacked patches of varying size (Fig. 5.4) in order to find whether or not they can match the required difference of phase delay. For the following results, the patches in each layer of dimensions $a_i \times b_i$ ($i = 1, 2, 3$) are assumed in a 14 × 14 mm square lattice, printed on a kapton film with 0.1-mm kevlar skins on both sides, and separated by Kevlar honeycomb of thickness t.

The phase of the reflection coefficient is computed at different frequencies by using the technique described in Chapter 3, based on Method of Moments and considering an infinite array. Fig. 5.16 shows the phase of the reflection coefficient at the central frequency (12 GHz) and the difference of phases $\boldsymbol{D}_r(f_1, f_2)$ at extreme frequencies (11.4 GHz, 12.6 GHz), versus the patch size for a two- and three-layer periodic structure, for normal incidence. For compensating the phase delay in the 1.5-m antenna for a 10 percent bandwidth, the reflection phase $\boldsymbol{\phi}_r(f_0)$ in Fig. 5.16(a) should match the phase-shift $\boldsymbol{\phi}_{oi}$ in Fig. 5.8 at 12 GHz on each element, and at the same time the difference $\boldsymbol{D}_r(f_1, f_2)$ given in Fig. 5.16(b) should match the values of $D_{di}(f_1, f_2)$ given in Fig. 5.15. Note that all possible combinations of phase-shift and phase difference are required, because phase difference increases from the edge to the center of the reflectarray (Fig. 5.15), while phase-shift varies very rapidly in quasi-concentric circles (Fig. 5.8).

On the other hand, the phase difference $\boldsymbol{D}_r(f_1, f_2)$ computed for the patch elements is always small for small size patches and increases with the patch dimensions (Fig. 5.16(b)). The phase difference $\boldsymbol{D}_r(f_1, f_2)$ can be controlled for

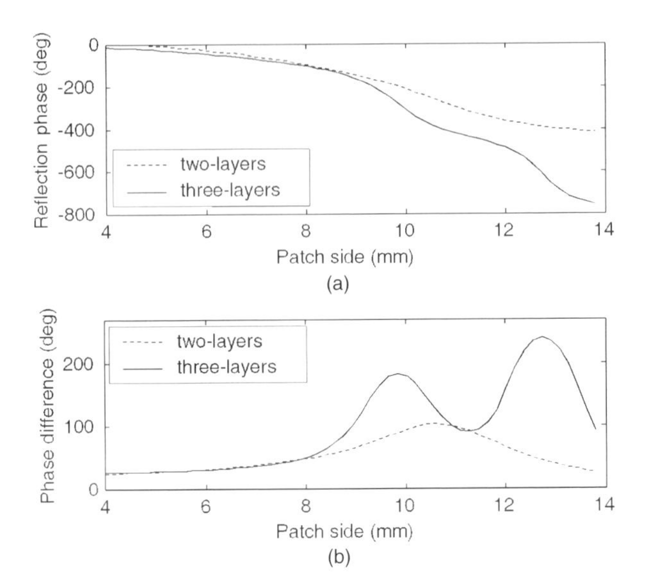

Figure 5.16. Reflection coefficient for a multilayer periodic structure with square patches for frequencies $f_0 = 12\,\text{GHz}$, $f_1 = 11.4\,\text{GHz}$, and $f_2 = 12.6\,\text{GHz}$. ($a_1 = 0.7a_2$, $t = 3\,\text{mm}$, for two array layers, and $a_1 = 0.7a_3$, $a_2 = 0.9a_3$, $t = 2\,\text{mm}$, for three array layers). (a) Phase, (b) phase difference.

large patches by adjusting the relative dimensions of the stacked patches, but not for small patches. When using two stacked arrays, $\boldsymbol{D_r}(f_1, f_2)$ will always be low for small and moderate values of the required phase-shift ϕ_{oi}, and then it cannot be adjusted to the required value of $\boldsymbol{D_{di}}(f_1, f_2)$. For this reason, a three-layer configuration must be chosen, which provides a range of phases around two times 360°, and any value of $\boldsymbol{\phi_{oi}}$ can be achieved with two different sizes of patches, giving more degrees of freedom to control the phase difference $\boldsymbol{D_r}(f_1, f_2)$.

The design for broadband operation is made by matching simultaneously the phase-shift at the central frequency $\boldsymbol{\phi_{oi}}$ and phase delay difference $\boldsymbol{D_{di}}(f_1, f_2)$ at each element (i) for the two linear polarizations. First, assuming a fixed relative size of the stacked patches, the two dimensions of the rectangular patches are adjusted in each cell to match the phase-shift at the central frequency by a zero searching routine, that iteratively calls an analysis routine based on the Method of Moments, as described in Section 3.5 of Chapter 3. The dimensions obtained from the design at the central frequency are used as starting point for the bandwidth optimization. To match $\boldsymbol{\phi_{oi}}$ and $\boldsymbol{D_{di}}(f_1, f_2)$ simultaneously, all the dimensions of the stacked patches are adjusted by using an optimization routine based on the Fletcher Powell algorithm, that minimize an error function defined as

$$E_i(a_1, b_1, a_2, b_2, a_3, b_3) = \sum_{l=1}^{2} (C_1(\phi_{ri}(f_0) - \phi_{oi})^2 + C_2(D_{ri}(f_1, f_2) - D_{di}(f_1, f_2))^2), \tag{5.4}$$

where C_1 and C_2 are weighting coefficients and l a subindex to indicate each linear polarization. Following the previous optimization process, an elliptical reflectarray with axes 1050 × 994 (mm) has been designed for dual polarization, to radiate a focused beam in the direction $\theta_b = 18°$, $\varphi_b = 0°$ in the 11.4–12.6 GHz band. A feed with a $\cos^4(\theta)$ field pattern has been assumed, that is, a taper illumination with −10 dB at the border, and its phase center located at coordinates $x_f = -334$, $y_f = 0$, $z_f = 900$ (mm) in the reflectarray coordinate system.

The radiation patterns on the x-z plane for x-polarization at the central and extreme frequencies are shown in Fig. 5.17 before and after optimization. Similar radiation patterns are obtained for the orthogonal linear polarization. For the nonoptimized reflectarray, the radiation patterns start to deteriorate in the band, with a 6-dB gain reduction, while a very stable gain was achieved for the optimized reflectarray, with only a 0.5 dB gain reduction. The optimization process compensates the narrowband effects produced by the element, and by the different path lengths. These results show that important bandwidth improvement and gain stability can be achieved for large reflectarrays by using optimization routines to accomplish the appropriate phase distribution within a frequency band.

For a bandwidth larger than 10 percent, matching the difference of phase at extreme frequencies is not sufficient to ensure a good performance in the whole frequency band, and the phase delay must be compensated at several frequencies in the band. Since the optimization to minimize the error function

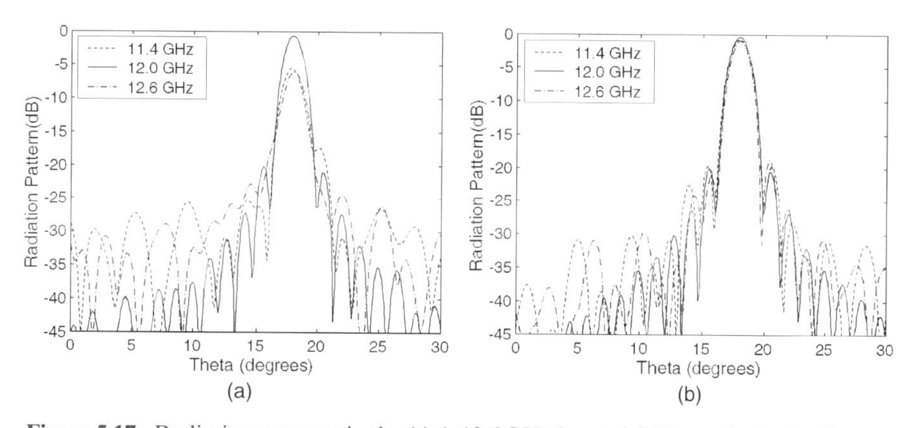

Figure 5.17. Radiation patterns in the 11.4–12.6 GHz band. (a) Nonoptimized reflectarray, (b) optimized reflectarray.

(5.4) involves six variables, the three patch lengths for each polarization, and the phasing performance for the two polarizations is practically uncoupled, the CPU time is significantly reduced if the optimization is carried out independently for each polarization. Then, for ***x***-polarization, only dimensions a_i are optimized to minimize the following error function

$$E_{xi}(a_1, a_2, a_3) = C_1(\phi^x_{ri}(f_0) - \phi^x_{oi})^2 + \sum_{l=1}^{N}(C_2(D^x_{ri}(f_l, f_0) - D^x_{di}(f_l, f_0))^2), \quad (5.5)$$

where l is a subindex to indicate each frequency in the band. Similarly, for **y**-polarization, the error function is

$$E_{yi}(b_1, b_2, b_3) = C_1(\phi^y_{ri}(f_0) - \phi^y_{oi})^2 + \sum_{l=1}^{N}(C_2(D^y_{ri}(f_l, f_0) - D^y_{di}(f_l, f_0))^2). \quad (5.6)$$

Note that for each polarization, the orthogonal dimensions of the patches are kept unchanged. After individual optimizations have been completed for each polarization, a simultaneous optimization is carried out for a fine adjustment of all patch dimensions, to take into account the slight influence of the orthogonal dimensions of the patches. This technique has been applied for the design of contoured beam reflectarrays in Ku-band of 1-m aperture, for 10 percent [25] and 15 percent bandwidth [26], and the results will be presented in Chapter 7.

5.4.3 Multifacet Reflectarrays

A multipanel configuration was proposed in [27] to reduce the errors associated with differential spatial phase delay. The configuration is made of several flat reflectarray panels arranged to approximate a parabolic surface. Fig. 5.18 shows three possible configurations from the simplest one-dimension arrangement to a more complex circumferential structure, similar to that of an umbrella. Each planar facet is a reflectarray that is designed to introduce the required phase-shift to simulate the real parabolic surface. Then, the compensation of phase delay in each panel should be approximately proportional to the distance d_{rp} from the flat panel to the ideal parabolic surface ($-2K_0 d_{rp}$). The panel dimensions can be reduced so that the required phase-delay in each panel is limited to a single 360° cycle. In this case, the reflectarray bandwidth will be limited only by the element behavior, but the complexity of manufacture is increased, as in the case of the circumferential arrangement shown in Fig. 5.18. Nevertheless, the multifacet configuration is totally compatible with the previous techniques for compensating the phase delay in a limited bandwidth, so the number of panels can be reduced in order to simplify the

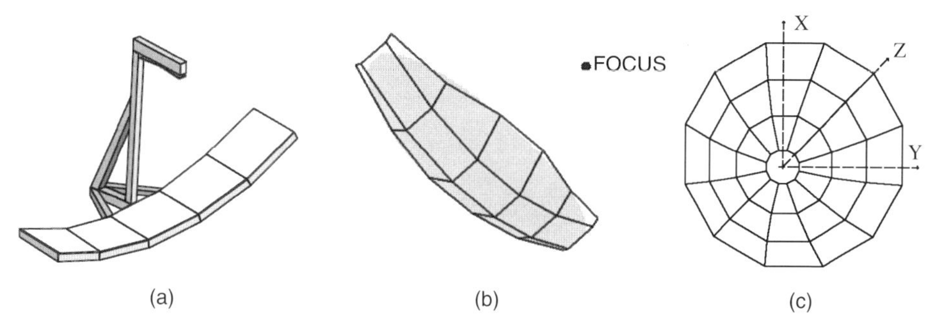

Figure 5.18. Multi-facet configuiration approximating a parabolic surface. (a) 1-D configuration, (b) rectangular-based arrangement, (c) circumferential arrangement.

manufacture or deployment of the antenna (Fig. 5.18 (a) and (b)) and the phase delay can be compensated in a range larger than 360° on each panel.

Although the multipanel configurations are more complex than a single flat reflectarray, they still offer the possibility of being folded and deployed for large antenna apertures. For example, the deployment techniques for the 1-D configuration will be similar as in solar arrays. For the rectangular-based configuration, in which the central panels are rectangular and the side panels trapezoidal, the deployment can be implemented by using pantographs, bracing strings and tensioned cables for achieving the required multifacet shape [24] (Fig. 5.19).

The simplest 1-D multipanel approach is valid for applications that require a large size in only one dimension, as was recently used for the NASA/JPL's Wide Swath Ocean Altimeter (WSOA) [28]. In this application, the required aperture, 2.16 m × 0.35 m, is made of five reflectarray panels, which make up a piecewise planar approximation of a parabola with a 1.125 m focal length, where the elements are varying-sized patches. Another suitable application for 1-D faceted reflectarray is found in Synthetic Aperture Radar (SAR) antennas. A three-panel configuration was proposed for an X-band SAR (9.65 GHz with 300 MHz bandwidth) [29]. The aperture was defined as 6 m × 1.6 m and the focal length as 2 m. The antenna is made of three panels, as shown in Fig. 5.20, where the length of the central and lateral panels is 1827 mm and 2331 mm, respectively. After computing the phase-shift required on each panel to produce the same beam at central frequency (9.65 GHz) as the reference parabolic reflector, and assuming the same phase-shift for the extreme frequencies, the radiation patterns are shown in Fig. 5.21 for azimuth and elevation planes. The effect of differential spatial phase delay in each panel produces an increase of the sidelobe level, particularly in the azimuth plane. This negative effect can be compensated by using a three-layer configuration with varying-sized patches

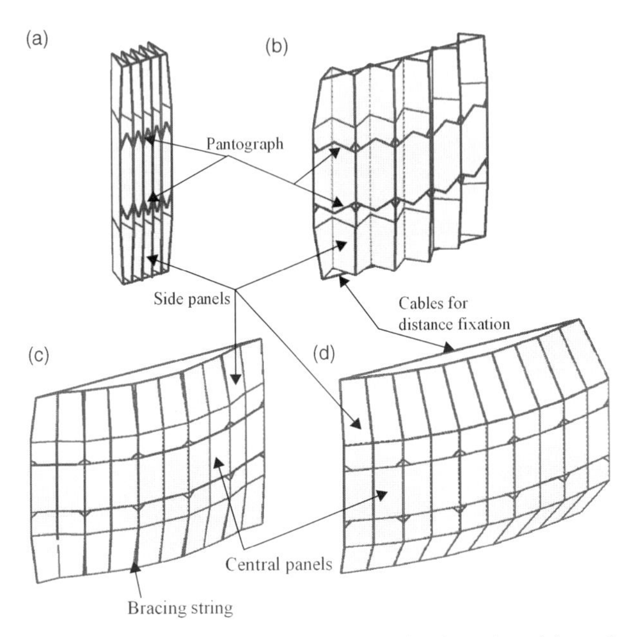

Figure 5.19. Deployment sequence for a rectangular based multipanel reflectarray.

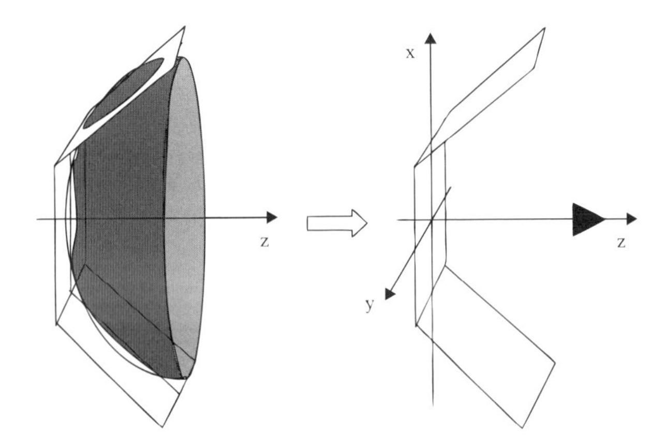

Figure 5.20. Three-panel centered-fed configuration.

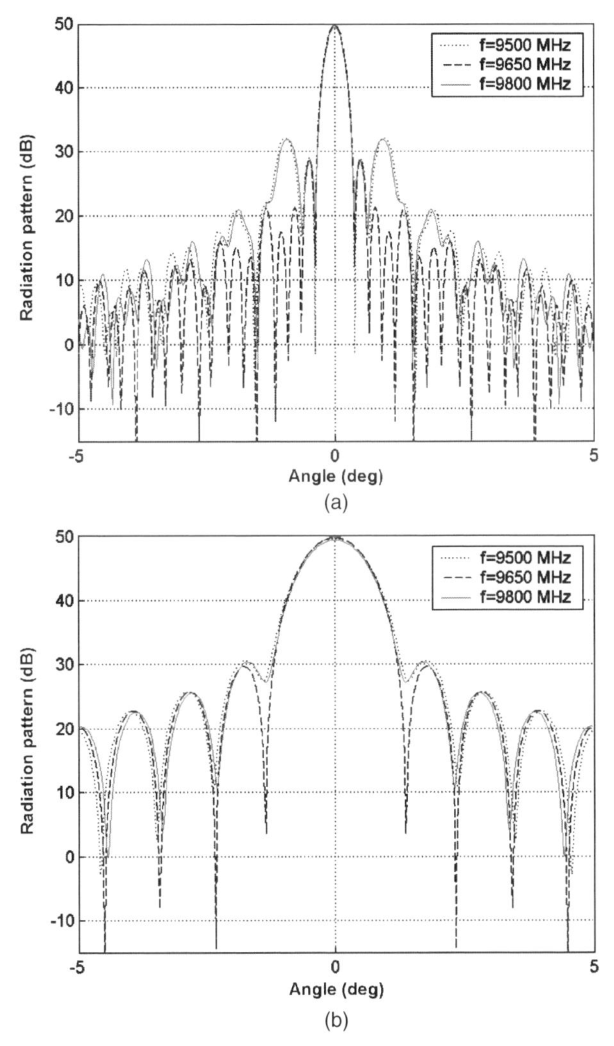

Figure 5.21. Radiation patterns of the three-panel antenna assuming the same phase-shift at the three frequencies. (a) Azimuth plane, (b) elevation plane.

and compensating the phase delay in the required frequency band (9.5–9.8 GHz), as described in the previous section. Each reflectarray panel was defined with the arrays of rectangular patches printed on Kapton, bonded to Kevlar sheets and separated by 3-mm thick Nomex-honeycomb. The x and y dimensions of the three stacked patches were optimized to match the required phase-shift at 9.5, 9.65, and 9.8 GHz for both linear polarizations at each element of the three panels. The resulting radiation patterns obtained by the analysis routine based on Method of Moments and local periodicity are shown

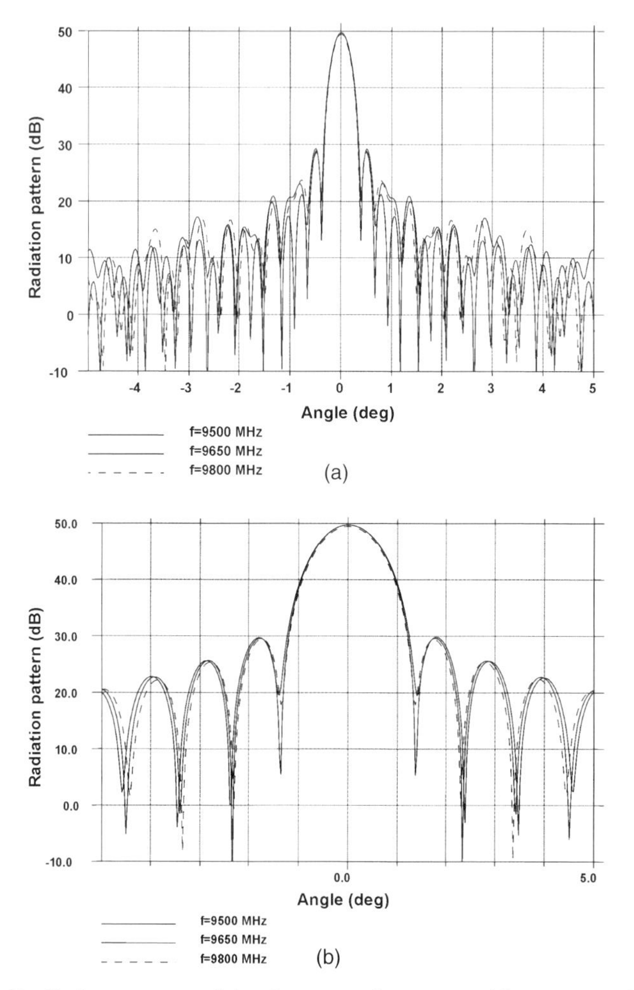

Figure 5.22. Radiation patterns of the three-panel antenna with compensation of phase delay on each panel. a) Azimuth plane, (b) elevation plane.

in Fig. 5.22, which show that the side lobes at extreme frequencies in Fig. 5.21 have been eliminated. These results show that the multifacet configuration can be used in conjunction with other proposed techniques for bandwidth improvement, particularly for very large apertures.

REFERENCES

1. J. Huang, "Bandwidth study of microstrip reflectarray and a novel phased reflectarray concept," in Antennas and Propagation Society Intl. Symp. 1995. AP-S. Digest, pp. 582–585, June 1995.

2. D. M. Pozar, "Bandwidth of reflectarrays," *Electr. Lett.* Vol. 39, No. 21, pp. 1490–1491 Oct. 2003.
3. A. W. Robinson, M. E. Bialkowski and H. J. Song, "An X-band passive reflect-array using dual-feed aperture-coupled patch antennas," *Asia Pacific Microwave Conf.*, pp. 906–909, Dec. 1999.
4. E. Carrasco, B. Alfageme, and J. A. Encinar, "Design of a multilayer aperture-coupled cell used as phase-shifter in reflectarrays," *Journées Internationales de Nice sur les Antennes*, Nice, France, November 2004.
5. E. Carrasco and J. A. Encinar, "Aperture-coupled patch reflectarray unit cell with a large range of phase delay," 28th ESA Antenna Workshop on Space Antenna Systems and technologies Noordwijk (The Netherlands), 30 May–3 June, 2005.
6. E. Carrasco, M. Barba, and J.A. Encinar, "Aperture-coupled reflectarray element with wide range of phase delay," *Electronics Letters*. Vol. 42, No. 12, pp. 667–668, June 2006.
7. E. Carrasco, M. Barba, and J. A. Encinar, "Reflectarray element based on aperture-coupled patches with slots and lines of variable length," *IEEE Trans. Antennas Propagat.*, Vol. 55, No. 3, pp. 820–825, March 2007.
8. J. A. Encinar, "Printed circuit technology multi-layer planar reflector and method for the design thereof," European Patent EP 1120856, June 1999.
9. J. A. Encinar, "Design of Two-Layer Printed Reflectarrays Using Patches of Variable Size," *IEEE Trans. Antennas Propagat.*, Vol. 49, No. 10, pp. 1403–14010, October 2001.
10. J. A. Encinar and J. A. Zornoza, "Broadband design of three-layer printed reflectarrays," *IEEE Trans. Antennas Propagat.*, Vol. 51, No. 7, pp. 1662–1664, July 2003.
11. M. Patel and J. Thraves, "Design and development of a low cost, electronically steerable, X-band reflectarray using planar dipoles," in *Proc. of Military Microwaves*, London, UK., pp. 174–179, 1994.
12. D. C. Chang and M. C. Huang, "Multiple-polarization microstrip reflectarray antenna with high efficiency and low cross-polarization," *IEEE Trans. Antennas Propagat.*, Vol. 43, pp. 829–834, Aug. 1995.
13. M. Bozzi, S. Germani, and L. Perregrini, "Performance comparison of different element shapes used in printed reflectarrays," *Antennas and Wireless Propagation Letters*, Volume 2, Issue 1, 2003 pp. 219–222.
14. N. Misran, R. Cahill, and V. Fusco, "Reflection phase response of microstrip stacked ring elements," *Electronics Letters*, Volume 38, Issue 8, pp. 356–357, April 2002.
15. N. Misran, R. Cahill, and V. Fusco, "Design optimisation of ring elements for broadband reflectarray antennas," Microwaves, Antennas and Propagation, *IEE Proceedings*, Volume 150, Issue 6, pp. 440–444, Dec. 2003.
16. M. R. Chaharmir, J. Shaker, M. Cuhaci, and A. Ittipiboon, "Broadband reflectarray antenna with double cross loops," *Electronics Letters*, Volume 42, Issue 2, pp. 65–66, Jan. 2006.
17. T. N. Chang and Y. C. Wei, "Proximity-coupled microstrip reflectarray," *IEEE Trans. Antennas and Propagat.*, Volume 52, Issue 2, pp. 631–635, Feb. 2004.
18. D. M. Pozar, "Comments on 'Proximity-coupled microstrip reflectarray'," *IEEE Trans. Antennas and Propagat.*, Volume 52, Issue 12, pp. 3411, Dec. 2004.

19. T. N. Chang and H. Suchen, "Microstrip reflectarray with QUAD-EMC element," *IEEE Trans. Antennas and Propagat.*, Volume 53, Issue 6, pp. 1993–1997, June 2005.
20. D. Cadoret, A. Laisne, R. Gillard, and H. Legay, "New reflectarray cells using coupled microstrip patches loaded with slot," Intl. Symposium on Antennas (JINA 2004), Nice, France, Nov. 2004.
21. D. Cadoret, A. Laisne, R. Gillard, L. Le Coq, and H. Legay, "Design and measurement of new reflectarray antenna using microstrip patches loaded with slot," *Electronics Letters*, Volume 41, Issue 11, pp. 623–624, May 2005.
22. S. M. Duffy and S. D. Targonski, "Comparison of two flat reflector-type designs for dual-polarization, dual-band operation," Antennas and Propagation Society Intl. Symp. pp. 288–291, July 2001.
23. S. Silver, *Microwave Antenna Theory and Design*, McGraw-Hill: New York, 1949.
24. J. A. Encinar, et al., "Reflectarray Antenna in Flat Panels (RAFP)," ESA TRP activity AO/1-4133/02/NL/JA, Final Report.
25. J. A. Encinar, L. Datashvili, J. Agustín Zornoza, M. Arrebola, M. Sierra-Castañer, J. L. Besada, H. Baier, and H. Legay "Dual-polarization dual-coverage reflectarray for space applications," *IEEE Trans. on Antennas and Propag.*, Vol. 54, pp. 2827–2837, Oct. 2006.
26. J. A. Encinar, M. Arrebola, M. Dejus, and C. Jouve, "Design of a 1-metre reflectarray for DBS application with 15% bandwidth," European Conference on Antennas and Propagation (EuCAP 2006), 6–10 November 2006, Nice, France.
27. A. Roederer, US6411255 "Reflector antenna comprising a plurality of panels," 2002-06-25.
28. R. Hodges and M. Zawadzki, "Wide swath ocean altimeter antenna electrical subsystem—preliminary design review," JPL internal document, January 2003.
29. H. Legay, B. Salome, E. Girard, S. Ramongassie, J. Encinar, and G. Toso, "Reflectarrays antennas for Sar missions," Antem 2005, 11th Intl. Symp. on Antenna Technology and Applied Electromagnetics, Saint-Malo, France, June 15–17, 2005.

CHAPTER 6

Dual-Band Reflectarray

The reflectarray antenna is well known for its narrow-bandwidth characteristics as explained in previous chapters. Nominally, its bandwidth varies from a few percent to about ten percent, depending on its f/D ratio and aperture electrical size [1–3]. By using wide-band elements or multilayer elements, the bandwidth of a reflectarray has currently been increased to 15 percent or more [4]. In the case where the system requires two widely separated frequencies, such as X- and Ka-bands, to share the same aperture, the conventional reflectarray fails to provide coverage. A parabolic reflector antenna, on the other hand, due to its specifically curved surface, can reflect unlimited frequencies and still achieve a planar phase front for all these frequencies. For a reflectarray, although it is limited in bandwidth, there are specific techniques to provide coverage of two widely separated frequencies. It is the purpose of this chapter to discuss several dual-band techniques for the reflectarray. Techniques to achieve more than two bands will also be briefly discussed.

6.1 DUAL-BAND WITH A SINGLE-LAYER SUBSTRATE

A single reflectarray to cover two-frequency bands with only a single-layer substrate is possible by using elements with structures having small widths, such as circular or square loops or crossed dipoles for circular or dual-linear polarization and just dipoles for linear polarization. With these techniques, a single-layer or a single membrane substrate can support all the dual-band elements to achieve a relatively simple antenna structure with lower weight. Depending on the polarization types, these techniques are separately discussed below.

6.1.1 Circular or Dual-Linear Polarization with Two Widely Separated Frequencies

For circular or dual-linear polarization using the variable-length delay line technique [5] to achieve phase compensation, multiple small square loops

Reflectarray Antennas, by John Huang and José A. Encinar

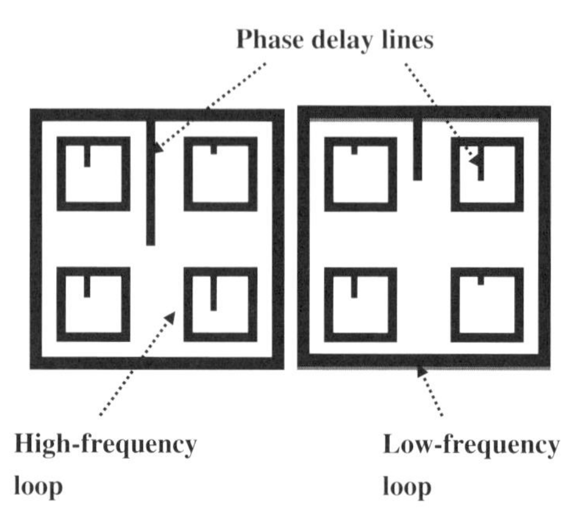

Figure 6.1. Dual-band elements using square loops with variable-length phase delay lines for CP or dual-linear application.

enclosed by a large square loop is a possible concept. This is illustrated in Fig. 6.1, where 2×2 small square loops are enclosed by a large square loop. This allows one frequency to be nearly twice the other frequency. More small loops can be enclosed in a large loop to achieve triple or quadruple frequency ratios. However, one must pay attention to the spacing between elements and their common ground plane. For the smaller elements, this spacing may become prohibitively large. As a consequence, small reflectarray elements with large spacing from the ground plane may result in a narrow element pattern, which will behave poorly for waves with low incident angles as explained in Chapter 4.

Another possible configuration is to use variable-size crossed dipoles [6], as illustrated in Fig. 6.2, where the smaller crossed dipoles will operate at a frequency about twice that of the larger crossed dipoles, such as 16 GHz and 8 GHz, respectively. One should notice that the previous loop-type of elements (square or circular) cannot be used with the variable-size technique for the low-frequency elements. This is because, as the elements change sizes, the low-frequency loops will most likely interfere physically with the high-frequency elements. On the other hand, by using crossed dipoles, the high-frequency elements can be replaced by either variable-size loops or variable-size patches [7], such as that illustrated in Fig. 6.3.

6.1.2 Circular Polarization with Two Closely Separated Frequencies

When two frequencies are fairly close to each other, such as 7 GHz and 8.5 GHz, two concentric circular loop elements can be used on the same substrate layer, as illustrated in Fig. 6.4. The phase compensations are achieved by using the variable angular rotation technique [8]. Variable-size technique cannot be used here; otherwise, physical interference between elements of the

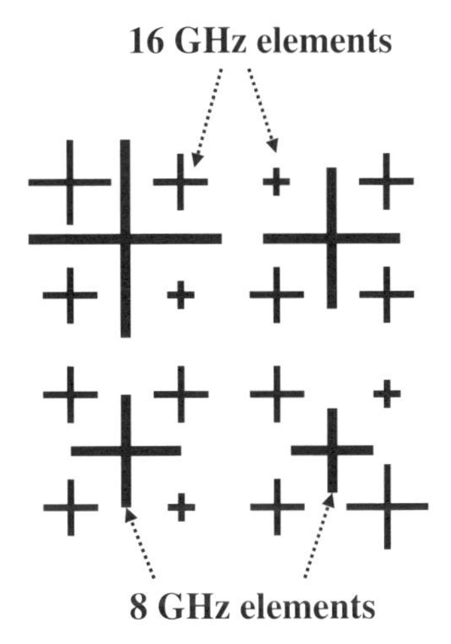

Figure 6.2. Dual-band elements using crossed dipoles with variable sizes for CP or dual-linear application.

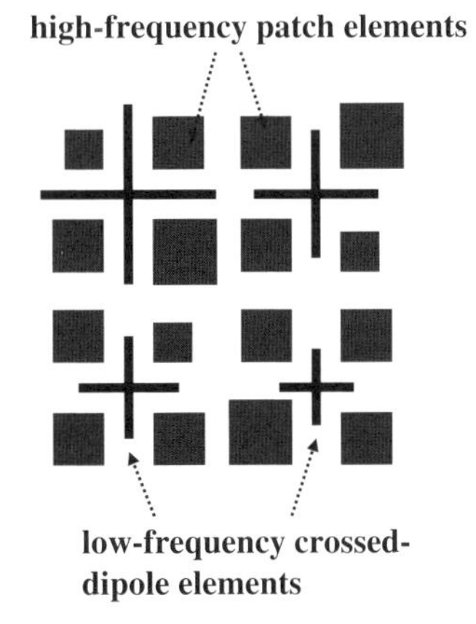

Figure 6.3. Dual-band elements using variable-sized crossed dipoles and variable-sized patches for CP and dual-linear application.

two bands will occur. The small gaps in each loop are capacitances, which make the loop element circularly polarized. Each element must be circularly polarized; otherwise, the variable angular rotation technique will not function properly.

6.1.3 Linear Polarization with Two Widely Separated Frequencies

When two or three frequencies are widely separated or nearly multiples of each other, linearly polarized dipoles of different lengths can be used similar

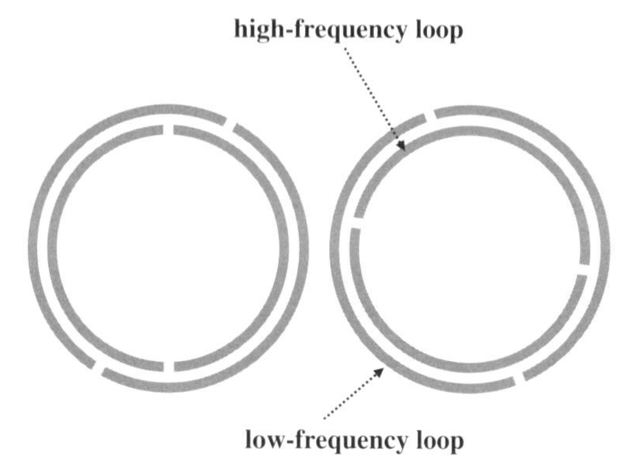

Figure 6.4. Dual-band elements using variable-angular rotated circular loops for CP application.

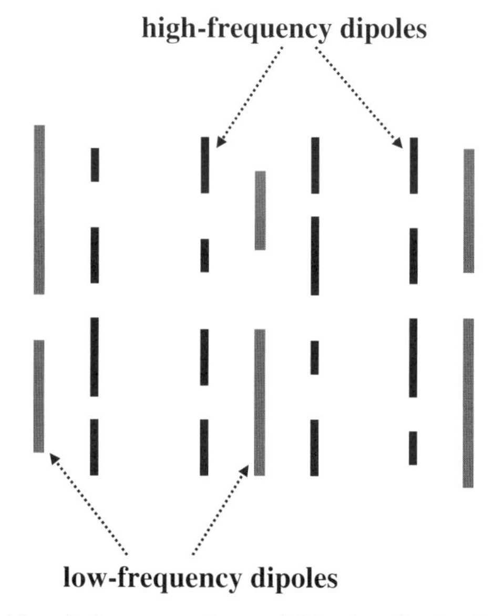

Figure 6.5. Dual-band elements using variable-size dipoles for LP application.

to that used for the crossed dipoles. The variable-size technique is used here for phase compensation. Fig. 6.5 gives an example of two widely separated frequencies using a single substrate with one frequency being twice that of the other.

Another approach is to use a rectangular patch with both of its orthogonal ports excited. Due to the two different dimensions of the rectangular patch, two different frequencies can be excited by using its orthogonal ports. Consequently, two orthogonal linear polarizations, each excited at a different frequency, can be formed from a single substrate reflectarray. Based on this concept, a reflectarray in orthogonal polarizations, operating at 24 GHz and 60 GHz, was demonstrated in [9]. Another reflectarray using Cassegrain geometry was also demonstrated in [10] for 19.4 GHz and 29 GHz.

6.1.4 Linear Polarization with Two Closely Separated Frequencies

Since a linear dipole occupies very little space when compared with a crossed-dipole, patch, or loop, many dipoles of different frequencies can be arrayed close to each other. Fig. 6.6 shows how dipoles of three different frequencies can be packed together with one frequency (f_1) being nearly half of the second frequency (f_2), while nearly equal to the third frequency (f_3). One should be cautioned that the elements should not be spaced too close to each other; otherwise, the mutual coupling effect may become too strong and significantly modify the phase-shift at both frequencies. Most up-to-date mathematical analysis tools use an infinite-array approach to predict the element's phase characteristics, where it assumes all neighboring elements are identical in size. Since variable-length dipoles are used and many other-frequency dipoles are present, the mutual coupling effect can only be predicted by the infinite array approach if the periodic cell includes the phasing dipoles of all other frequencies, as shown in Fig. 6.6.

6.2 DUAL-BAND WITH TWO-LAYER SUBSTRATES

When the two frequencies are as near to each other, as 6.5 GHz and 10.5 GHz, the same period can be used for both frequency bands, and two stacked patches with variable size can be used to independently adjust the phase at the two frequencies [11]. The reflectarray element is the same as shown in Fig. 3.13, but now the dimensions of the upper and lower patches are varied to adjust the phase at the high and low frequencies, respectively. A center-fed

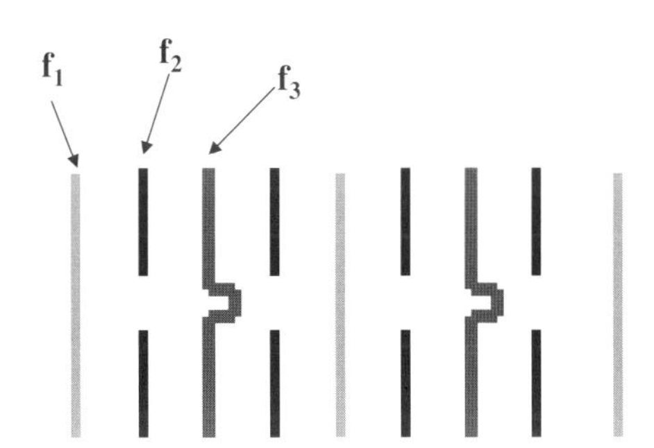

Figure 6.6. Tri-band elements using variable-size dipoles for LP Application; $f_1 \approx 0.5f_2$, $f_1 \approx f_3$.

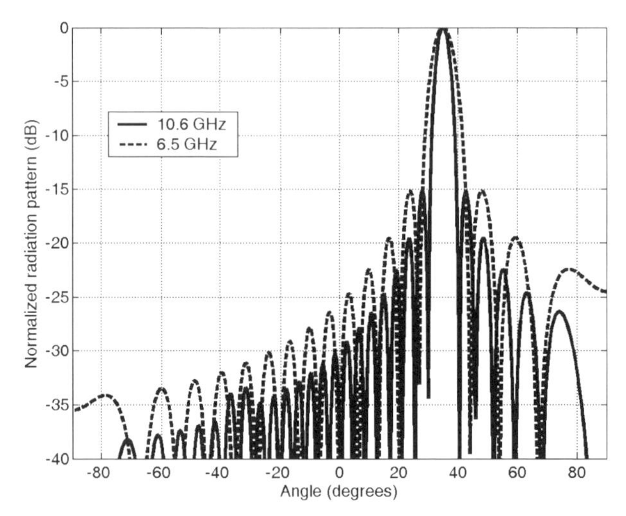

Figure 6.7. E-plane radiation pattern for a dual-frequency reflectarray based on stacked patches of variable size.

reflectarray with the feed placed at 35 cm has been designed to produce a beam in the direction $\theta_0 = 35^\circ$ in the XZ plane at 6.5 and 10.5 GHz. Fig. 6.7 shows the theoretical E-plane patterns at 6.5 and 10.5 GHz.

When two frequencies are widely separated, such as f_1 is nearly four times f_2, the physical separation between the elements and their common ground plane may become an issue as explained in Section 6.1.1. In this situation, two-layer substrates instead of a single-layer substrate may be a solution. There are two different approaches for the two-layer technique. One places the layer of high-frequency elements above the low-frequency elements, while the other places the low-frequency elements above the high-frequency elements. These two approaches are separately discussed below.

6.2.1 High-Frequency Elements above Low-Frequency Elements

Since the two frequencies are widely separated, the low-frequency elements are significantly larger than the high-frequency elements in physical size. As a result, a single low-frequency element can be used as a ground plane for several high-frequency elements. This approach was first reported in 1995 [12], where the variable-length phase-delay-line technique was used with a unit cell consisting of 2×2 high-frequency patches placed above a single low-frequency patch, as illustrated in Fig. 6.8. With this approach, the lower element must use a fixed-size patch with variable-length phase-delay-line technique; otherwise, if a variable size patch were used, some of the upper-layer elements may not see an appropriate ground plane below them. The upper elements, on the other hand, due to their smaller size, could use a variety of element types, such as

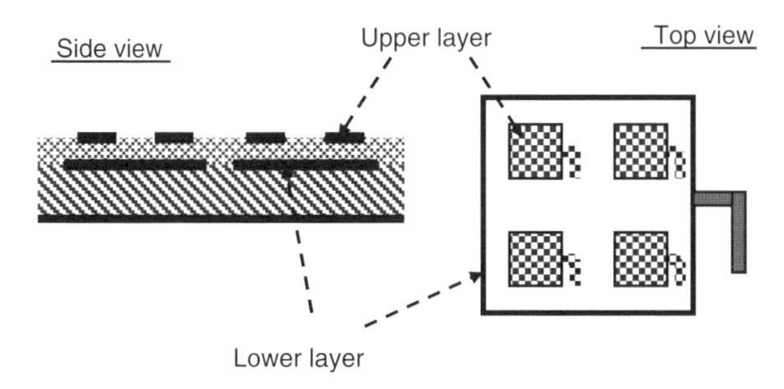

Figure 6.8. Configuration of a dual-band dual-layer reflectarray.

variable-sized crossed dipoles, variable-angular rotated loops (for CP only), etc. For the approach used in Fig. 6.8, the report in 1995 [12] used the Method of Moments to calculate the radar cross-section (RCS) of the two-layer structure. It did predict two resonant frequencies corresponding to the two different-size patches, which indicates that the structure in Fig. 6.8 will indeed function properly as a reflectarray at the two desired frequency bands.

6.2.2 Low-Frequency Elements above High-Frequency Elements

The large low-frequency elements can be placed above the small high-frequency elements only if they are electrically nearly transparent to the high-frequency waves. To do so, the low-frequency element has to use a structure with small widths, such as the crossed-dipole or annular ring loop. A solid patch cannot be used since it will completely block off the high-frequency waves. The large low-frequency element not only has to have small structure widths, but also has to avoid higher-mode resonances that occur at the integer multiple of the high-frequency element. In other words, the low-frequency layer must act as a transparent frequency selective surface (FSS) to the high-frequency layer situated below. There are two different up-to-date approaches that have been implemented as demonstration breadboards, and they are separately discussed below.

6.2.2.1 Low-Frequency Crossed-dipoles above High-Frequency Patches. For NASA/JPL's deep-space telecommunications, a high-gain, circularly polarized spacecraft antenna with a size above 3 m has always been needed with the capability of transmitting a downlink frequency at Ka-band of 32 GHz, while receiving an uplink signal at X-band of 8.4 GHz. A dual-band reflectarray would certainly be a candidate. To demonstrate the dual-band technology, a 0.5-m reflectarray was developed [13] with the above dual-band capability. Illustrated in Fig. 6.9 as a unit cell, the reflectarray used variable-sized X-band crossed-dipoles placed above the variable-size Ka-band patches, which is a similar approach as that configured in Fig. 6.3. The X-band

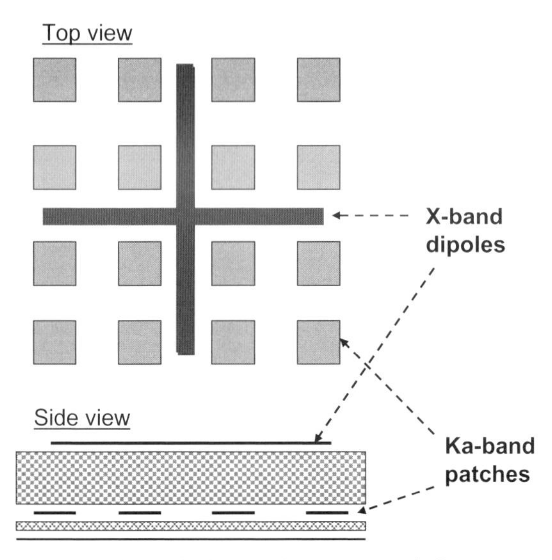

Figure 6.9. Reflectarray unit cell of an X-band crossed-dipole above 4×4 Ka-band patches.

crossed-dipoles are physically thin enough so as not to cause significant blockage to the Ka-band signal, while the Ka-band patches, with their relatively thin substrate, serve as part of the ground plane for the X-band elements. It happens that the Ka-band frequency of 32 GHz is approximately four times the frequency of the X-band frequency at 8.4 GHz, which, by properly designing the substrate's dielectric constant and element spacing, allows one X-band element to accommodate 4×4 Ka-band elements. With the X-band crossed-dipoles not situated directly above the Ka-band elements, the blockage effect of the X–band elements to the Ka-band signal should be minimized. The 0.5-m dual-band circularly polarized reflectarray antenna is shown in Fig. 6.10, where the top X-band layer is purposely shifted in location to expose the bottom Ka-band layer. The measured patterns of both X-band and Ka-band frequencies are given in Fig. 6.11. These patterns also show the effect of each frequency with and without the existence of the other frequency's layer. The existence of the Ka-band layer had negligible effect on the X-band performance, while the existence of the X-band layer did affect the sidelobe level and gain of the Ka-band signal. The Ka-band signal lost a gain of 1.8 dB with the presence of the X-band layer. Although both X- and Ka-band patterns are fairly good, with sidelobes below –20 dB, the measured antenna efficiencies are relatively poor, with numbers slightly below 30 percent. The expected efficiencies should be above 45 percent. It is believed that both the software (Method of Moments) in designing the antenna and the reflectarray configuration design need to be investigated and improved.

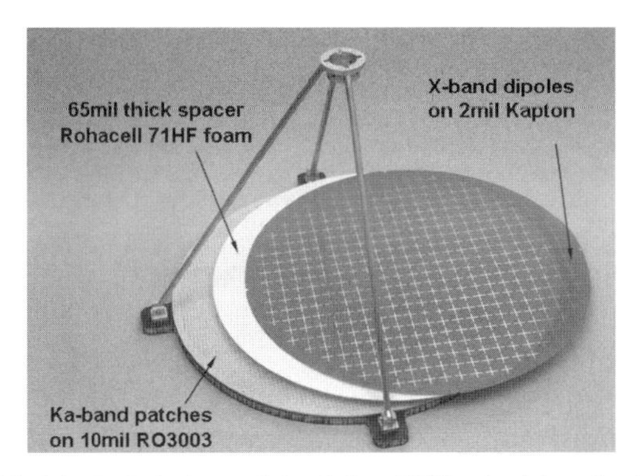

Figure 6.10. Exploded view of the 0.5-m X/Ka dual-band reflectarray.

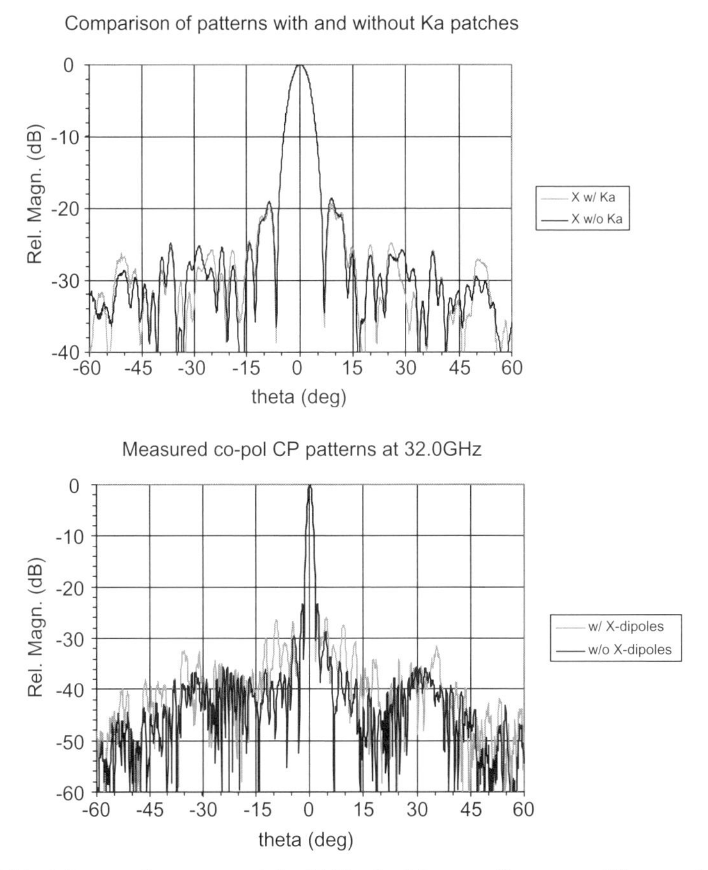

Figure 6.11. Measured patterns of the X/Ka dual-bang reflectarray. The top shows the X-band patterns with and without the Ka-band layer, the bottom shows the Ka-band patterns with and without the X-band layer.

6.2.2.2 Low-Frequency Annular Rings above High-Frequency Annular Rings. For the same NASA/JPL deep-space telecommunications requirement as discussed in the previous section, another approach to generate a circularly polarized high-gain reflectarray antenna is presented here. It also places low-frequency elements on top of high-frequency elements but uses annular ring elements [14] with variable angular rotation technique [8]. The annular ring element not only has a physically thin structure so it behaves nearly transparent to the other frequency, but also has a broader CP bandwidth relative to other types of elements. The broader CP bandwidth comes from a range of nearby resonating frequencies occurring due to the curvature of the ring structure. To utilize the angular rotation technique, two oppositely located small gaps are introduced onto the ring element structure so that the incoming single-sensed CP signal could recognize the rotation direction and reradiates with the same-sensed CP signal. In other words, by adding gaps to the ring structure and thus adding capacitance, the ring elements are transformed from a none-polarization-sense element to a single-sensed CP element. To demonstrate the dual-band technology, a series of small-size reflectarrays were developed. It started with a 0.5-m center-fed C/Ka dual-band reflectarray [15] and then a 0.5-m offset-fed X/Ka dual-band reflectarray [16]. Both antennas demonstrated excellent pattern, efficiency, and axial-ratio performance.

The more recent demonstration was a 0.75-m Cassegrain offset-fed X/Ka dual-band reflectarray built on thin membranes [17], which is discussed in more details below. This Cassegrain reflectarray is shown in Fig. 6.12 with its dual-layer membranes illustrated in Fig. 6.13. Each layer membrane has a thickness of 0.0508-mm and is used solely to support the ring elements. The primary

Figure 6.12. 0.75-m Cassegrain offset-fed X/Ka dual-band reflectarray mounted on antenna range pedestal.

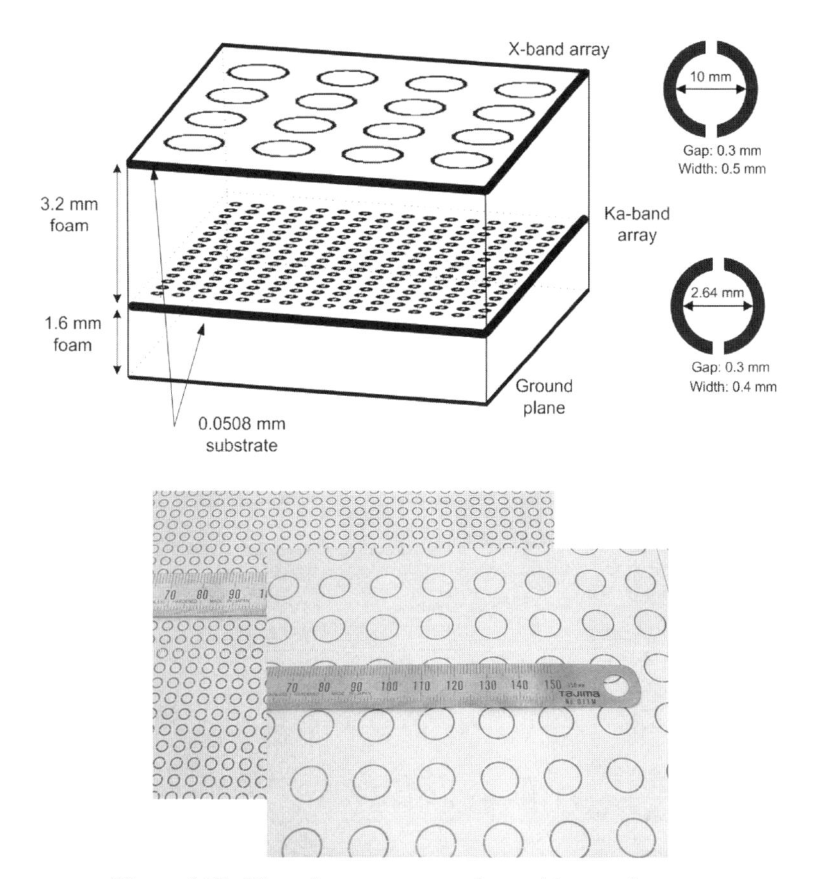

Figure 6.13. Ring elements on two-layer thin membranes.

substrates for the two bands are mostly air or empty space. However, for this technology demonstration, low-dielectric-constant foam layers are substituted for the empty spaces to act as supporting structures for the thin membranes.

In an actual space flight unit, an inflatable or rigid-frame structure will provid proper tensioning force to eliminate the use of foam layers. The subreflector, which is a convex hyperboloid, has a projected elliptical aperture with major axis dimension of 150 mm and minor axis dimension of 84 mm. The feed antenna eventually could be a dual-band horn whose technology has already been developed previously. To minimize this reflectarray development effort, two microstrip arrays (one at X-band, the other at Ka-band) with relatively simple development effort were used to feed the subreflector at two different times. Due to the offset geometry, the main beam of the reflectarray was designed to tilt at 30° from the broadside direction in the elevation plane. This would allow both the reradiated and the reflected component fields of the

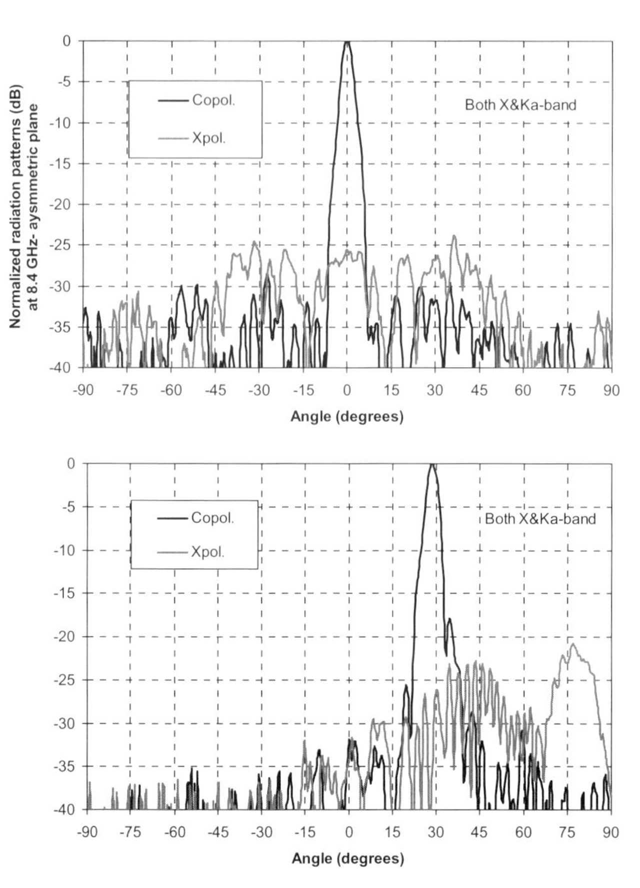

Figure 6.14. Measured two-principal-plane patterns at X-band for the Cassegrain offset-fed dual-band reflectarray.

reflectarray to travel in the same general direction and thus to achieve better efficiency.

The measured radiation patterns are given in Figs. 6.14 and 6.15, where they show excellent sidelobe and cross-polarization performances. In particular, the Ka-band patterns show cross-polarization levels well below −30 dB and sidelobe levels below −25 dB in all angular regions. The measured 3-dB axial-ratio bandwidth for X-band is more than 600 MHz (7.1 percent) and for Ka-band is more than 2 GHz (6.3 percent). It is worth mentioning here that the cross-polarization level of the reflectarray (<−30 dB) is better than the feed array (−25 dB). This is the result of using the angular rotation technique where the cross-polarization component of the feed is suppressed by the huge number of reflectarray elements augmented with the "sequential rotation" phenomenon [18, 19]. The measured overall antenna efficiency at X-band is 50 percent and at Ka-band is 48.2 percent. A comparison of the Ka-band performance

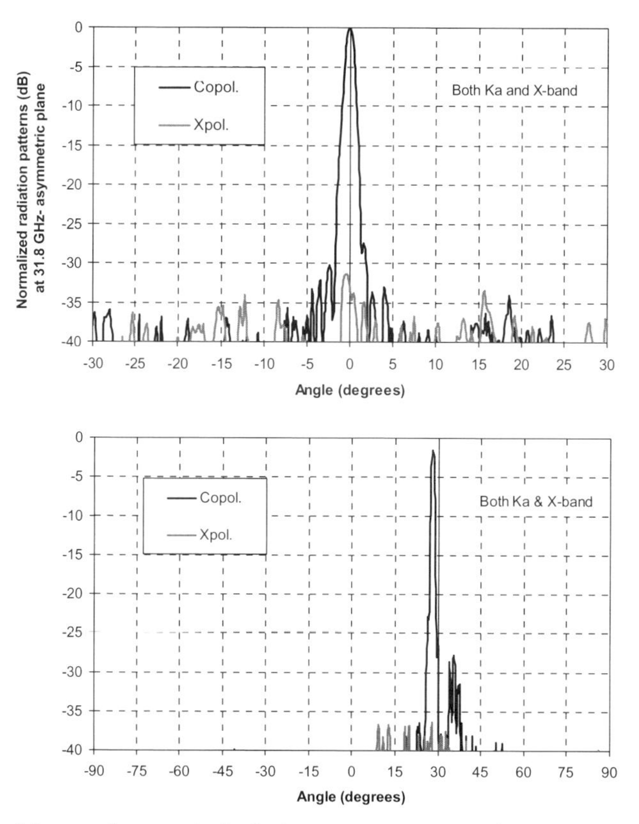

Figure 6.15. Measured two-principal-plane patterns at Ka-band for the Cassegrain offset-fed dual-band reflectarray.

with and without the X-band layer placed on top of the Ka-band layer was made. It showed that the presence of the X-band layer had negligible impact on the Ka-band pattern performance. However, the gain did suffer a 1 dB loss with the presence of the X-band layer, while the presence of the Ka-band layer slightly above the ground plane had negligible effect on the X-band performance in both pattern and gain.

6.3 MULTIBAND REFLECTARRAY WITH MORE THAN TWO FREQUENCIES

This chapter has presented information on mostly dual-band reflectarray designs. Although no development has been carried out for more than two frequency bands, a single reflectarray is conceptually feasible to provide such performance. Section 6.1.4 and Fig. 6.6 have illustrated a conceptual

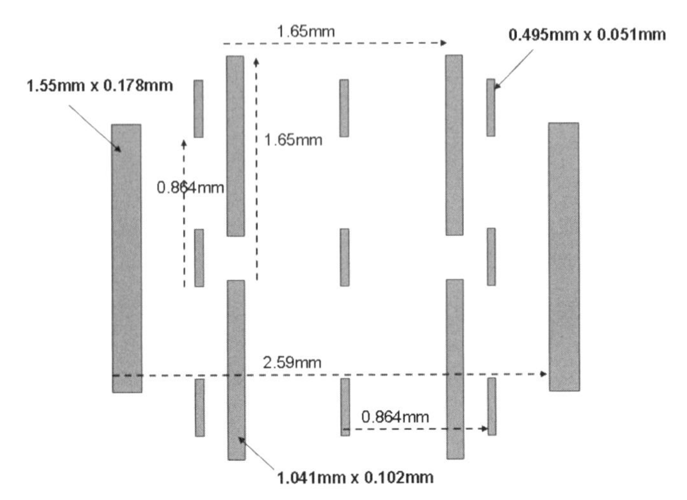

Figure 6.16. Three sets of printed dipole elements for resonant frequencies occurring at 60 GHz, 89 GHz, and 183 GHz.

design where three frequencies with linear polarization could be formed by using three different-length dipoles. An analysis using moment method techniques with Ansoft software has been performed to demonstrate such a possibility.

A radar cross-section (RCS) calculation was performed on three sets of different-length linearly polarized dipoles, as shown in Fig. 6.16. All these dipoles, shown together as a unit cell of a large reflectarray, are printed on a substrate of thickness 0.25 mm with relative dielectric constant of 2.9. One set of dipoles has dipole dimensions of 0.495 mm × 0.051 mm intended to resonate at 183 GHz, the second set has dipole dimensions of 1.041 mm × 0.102 mm intended for 89 GHz, while the last set has dipole dimensions of 1.55 mm × 0.178 mm designed for 60 GHz. The RCS result, calculated at the broadside direction, is given in Fig. 6.17, where it clearly shows three resonances with peaks occurring at the designed frequencies. The different amplitude peaks yield little consequence to the reflectarray performance. They are merely the indication of the amount of effective RCS aperture size. The 183-GHz peak is higher than the other two, because it may contain the higher-order-mode components of the other two dipole structures.

This is also indicated in the cross-polarization component, where the RCS keeps increasing toward the high-frequency end. By placing another thin layer of printed elements on top of this three-frequency layer, it is possible to have a reflectarray to achieve four or five frequency bands. For example, on the top layer, concentric square loops, as shown in Fig. 6.18, intended for either dual-linear polarization or circular polarization could be designed for lower frequency resonances at 6.6 GHz and 10 GHz, as shown in Fig. 6.19. This figure

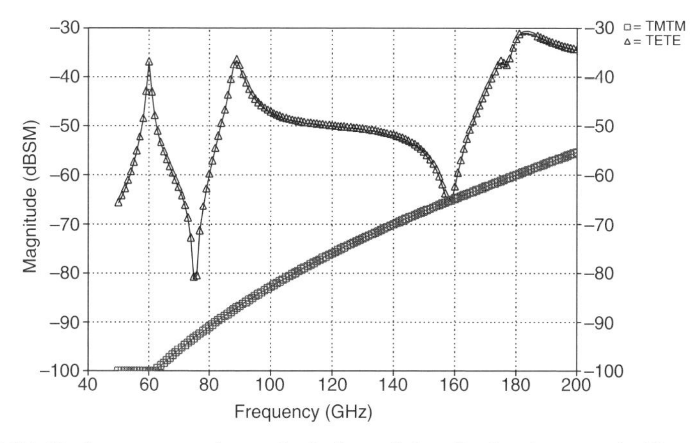

Figure 6.17. Radar cross-section calculation of the dipole elements in Figure 6.16. It shows three resonances at 60 GHz, 89 GHz, and 183 GHz.

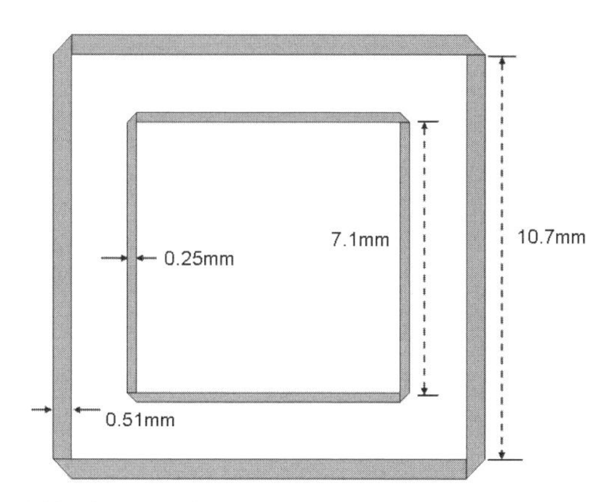

Figure 6.18. Concentric square loops for dual-band resonances.

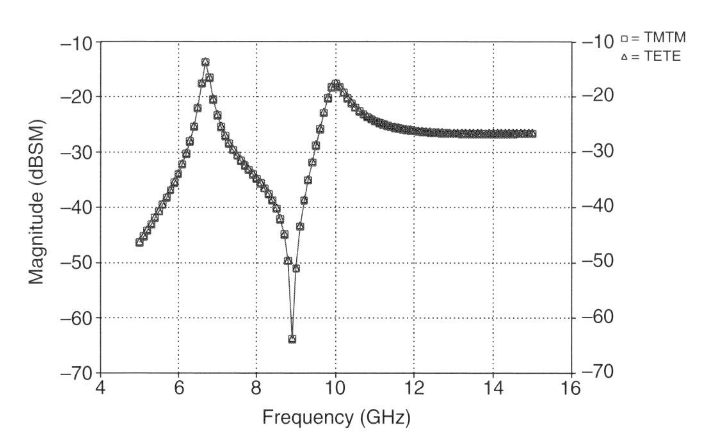

Figure 6.19. Radar cross-section calculation of the concentric square loops in Figure 6.17. It shows two resonances at 6.6 GHz and 10 GHz.

shows that, due the square structure, both vertical and horizontal polarizations have the same amount of RCS.

In this two-layer construction, the top-layer, low-frequency square loops act as a frequency selective surface to the lower-layer, high-frequency dipoles and allow these high-frequency signals to go through, while the lower-layer dipoles act as portion of the ground plane for the top-layer, low-frequency waves. As an overall result, a reflectarray with five frequencies is a possibility in the future.

REFERENCES

1. J. Huang, "Bandwidth study of microstrip reflectarray and a novel phased reflectarray concept," IEEE AP-S/URSI symposium, Newport Beach, California, June, 1995, pp. 582–585.
2. D. M. Pozar, "Bandwidth of reflectarrays," *Electronics Letters*, Vol. 39, Oct. 2003, pp. 1490–1491.
3. J. Huang, "Analysis of a microstrip reflectarray antenna for microspacecraft applications," JPL TDA Progress Report No. 42-120, February 15, 1995.
4. J. A. Encinar and J. A. Zornoza, "Broadband design of three-layer printed reflectarrays," *IEEE Trans. Antennas Propagat.*, vol. 51, July 2003, pp. 1662–1664.
5. J. Huang, "Microstrip reflectarray," IEEE AP-S/URSI symposium, London, Canada, June 1991, pp. 612–615.
6. A. Kelkar, "FLAPS: conformal phased reflecting surfaces," Proc. IEEE National Radar Conf., Los Angeles, California, March 1991, pp. 58–62.
7. D. M. Pozar and T. A. Metzler, "Analysis of a reflectarray antenna using microstrip patches of variable size," *Electronics Letters*, April 1993, pp. 657–658.
8. J. Huang and R. J. Pogorzelski, "A Ka-band microstrip reflectarray with elements having variable rotation angles," *IEEE Trans. Antennas Propagat.*, Vol. 46, May 1998, pp. 650–656.
9. D. Pilz and W. Menzel, "Printed millimeter-wave reflectarrays," *Annales des Telecommun.*, 56, No. 1–2, 2001, pp. 51–60.
10. J. Shaker, C. Pike, and M. Cuhaci, "A dual orthogonal Cassegrain flat reflector for Ka-band application," *Microwave and Optical Technology Letters*, Vol. 24, No. 1, Jan. 2000, pp. 7–10.
11. J. A. Encinar, "Design of a dual-frequency reflectarray using microstrip stacked patches of variable size," *Electronic Letters*, Vol. 32, No. 12, June 1996. pp. 1049–1050.
12. D. I. Wu, R. C. Hall, and J. Huang, "Dual-frequency microstrip reflectarray," IEEE AP-S/URSI symposium, 1995, pp. 2128–2131.
13. M. Zawadzki abd J. Huang, "A dual-band reflectarray for X- and Ka-bands," PIERS symposium, Honolulu, Hawaii, October 2003.
14. C. Han and K. Chang, "Ka-band reflectarray using ring elements," *Electron. Lett.*, Vol. 39, March 2003, pp. 491–493.

15. C. Han, C. Rodenbeck, J. Huang, and K. Chang, "A C/Ka dual-frequency dual-layer circularly polarized reflectarray antenna with microstrip ring elements," *IEEE Trans. Antennas Propagat.*, Vol. 52, Nov. 2004, pp. 2871–2876.
16. C. Han, J. Huang, and K. Chang, "A high efficiency offset-fed X/Ka dual-band reflectarray using thin membranes," *IEEE Trans. Antennas Propagat.*, Vol. 53, Sept. 2005, pp. 2792–2798.
17. J. Huang, C. Han, and K. Chang, "A Cassegrain offset-fed dual-band reflectarray," IEEE AP-S/URSI Symposium, Albuquerque, New Mexico, July 2006, pp. 2439–2442.
18. T. Chiba, Y. Suzuki, and N. Miyano, "Suppression of higher modes and cross polarized component for microstrip antenna," IEEE AP-S/URSI Symposium, May 1982, pp. 285–288.
19. J. Huang, "A technique for an array to generate circular polarization with linearly polarized elements," *IEEE Trans. Antennas Propagat.*, Vol. AP-34, Sept. 1986, pp. 1113–1124.

CHAPTER 7

Recent and Future Applications

The printed reflectarray antenna, as mentioned previously, combines many of the salient features of the traditional parabolic reflector antenna and microstrip array technologies. Its multitude of applications has just been discovered in the past century and will continue to be developed in the future. Many of the early and recent applications have been briefly discussed in Chapter 2, where the reflectarray's development history was narrated. This chapter gives a more detailed description of some of the recent and possible future applications, which are separately presented in the following sections. In particular, the authors were personally involved in the development of the applications in the first two sections, and, for this reason, the descriptions of these two applications are much more in depth.

7.1 INFLATABLE/THIN-MEMBRANE REFLECTARRAYS

The United States' NASA space science missions can be categorized into two major areas: deep-space exploration and Earth remote sensing. As much science information has been learned from previous missions, higher capabilities and more stringent system requirements are being placed on future missions, such as longer distance communication, higher data rates, finer radar imaging resolution, etc. Almost all these stringent requirements call for higher-gain and larger-aperture spacecraft antennas. At the same time, however, lower mass and smaller stowage volume for the spacecraft antenna are demanded in order to reduce payload weight and required shroud space, and thus minimize overall launch cost.

To meet these goals, several space deployable antenna concepts [1] have been investigated over the past several decades. To name a few, there were the Harris Corporation's hoop-column umbrella type, Lockheed's wrapped-rib version, TRW's Sun-flower antenna, and the more recent Astro mesh. All these deployable antennas are of the parabolic reflector type with metalized mesh reflecting surfaces. Because they have been parabolic with a relatively small

Reflectarray Antennas, by John Huang and José A. Encinar

focal length, they lack wide-angle beam scanning ability; only a few beamwidths can be scanned. The current technology of the mesh surface also limits the upper frequency of its operation to Ku-band or lower. In addition, some of these concepts suffer from higher risk because of too many mechanical components.

One good example of mechanical component failure is the Galileo spacecraft, which used the umbrella radial rib mesh reflector that failed to properly deploy in space. To remedy these drawbacks, several new antenna concepts are being investigated at the Jet Propulsion Laboratory (JPL) for possible future-mission applications. One of these concepts is the inflatable thin-membrane reflectarray. Two inflatable reflectarrays have been developed; one is at X-band with a 1-m diameter and the other is at Ka-band with a 3-m diameter. These are separately described below.

7.1.1 X-Band 1-m Inflatable Reflectarray

This antenna [2], shown in Fig. 7.1, has an inflated torus tube that supports and tensions the 1-m-diameter, two-layer-membrane reflectarray surface. The antenna's overall RF system and the aperture membrane surface were designed at JPL, while the inflatable structure and antenna integration were developed by ILC Dover, Inc. The inflated tripod tubes are attached to the torus as struts to support a feed horn. The inflatable tubes are made of 0.25-mm-thick urethane-coated Kevlar material, and the reflectarray membrane material is made of 5-micron-thick copper cladding on 0.05-mm-thick Kapton. There are a total

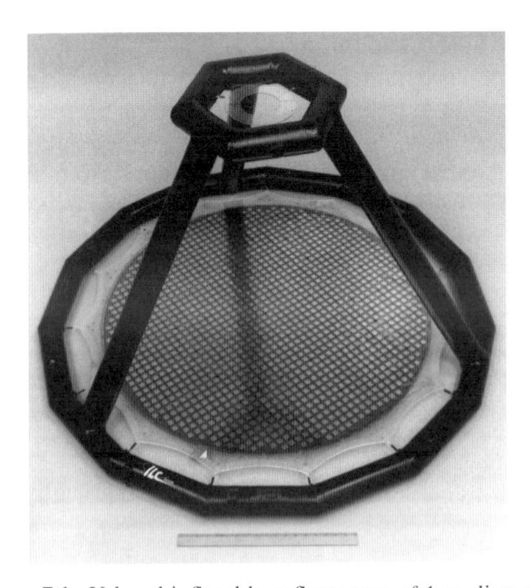

Figure 7.1. X-band inflatable reflectarray of 1-m diameter.

of two membrane layers separated by 1.3 mm. The top copper layer was etched to produce approximately 1000 isolated microstrip patches, while the unetched bottom layer serves as the ground plane. Many small foam discs (7-mm diameter) are placed between the two membranes as a means of maintaining the required uniform membrane spacing.

This inflatable antenna structure achieved a mass of 1.2 kg, which excludes the mass of the inflation system and the feed horn. The microstrip patches on the top layer membrane are all square in shape and identical in size, but with different-length phase-delay lines attached. Each patch has two equal-length phase-delay lines orthogonally attached. With the feed horn left-hand circularly polarized, the reradiated fields from the patches are right-hand circularly polarized. A measured antenna elevation pattern at 8.3 GHz is given in Fig. 7.2, where the peak sidelobe level (−19 dB) and peak cross-polarization level (−19 dB) are both acceptable to the communication system. However, the peak sidelobe of −19 dB is higher than the expected −25 dB. This is primarily due to the blockage effects of the feed and feed support struts. The main beam of the antenna has a −3 dB beamwidth of 2.4°, which is expected from a circular aperture of 1-m diameter. The antenna also achieved the expected −1 dB gain-bandwidth of 250 MHz (about 3 percent). The measured peak gain at 8.3 GHz is 33.7 dBi, which implies an antenna efficiency of 37 percent. The relatively large feed and strut blockage is the primary contributor to the inefficiency, which is believed to be correctable for future models.

7.1.2 Ka-Band 3-m Inflatable Reflectarray

This antenna [3], shown in Fig. 7.3, was co-developed by JPL and ILC Dover, Inc. It consists of a horse-shoe-shaped inflatable tube that supports and

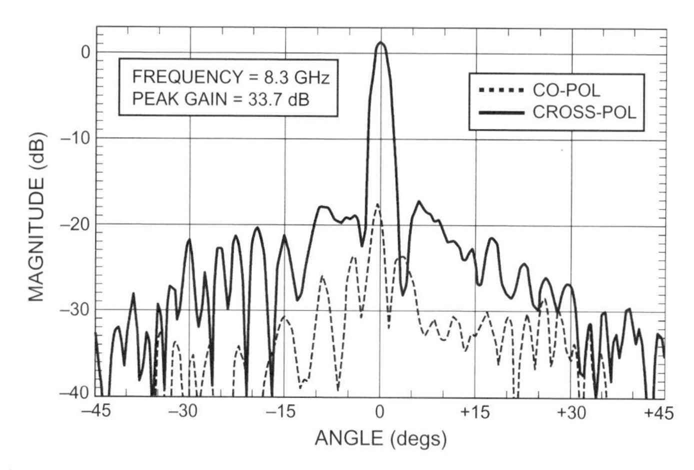

Figure 7.2. Measured radiation pattern of the X-band inflatable reflectarray shown in Fig. 7.1.

Figure 7.3. 3-m diameter Ka-band inflatable reflectarray.

tensions a 3-m aperture membrane. The tube, 25 cm in diameter, is made of urethane-coated Kevlar and is inflated to 3.0 pounds-per-square-inch (psi) pressure, which translates to about 90 psi of tension force to the aperture membrane. The inflatable tube is connected to the aperture membrane at 16 catenary points with spring-loaded tension cords. Each connecting point has displacement adjustment capability in the x, y, z directions so that the circumference of the circular aperture membrane can be made into a single plane orthogonal to the feed horn axis. The single-layer aperture membrane is a 5-mil (0.13 mm) thick Uplex™ dielectric material (a brand of polyimide), with both sides clad with 5-micron-thick copper. The copper on one side was etched to form approximately 200,000 microstrip patch elements, while the copper on the other side is unetched and serves as the ground plane for the patch elements. A portion of the microstrip elements is shown in Fig. 7.4. The elements use the variable rotation technique [4] to provide the needed electrical phases. The inflatable tripod tubes, asymmetrically located on the top portion of the horse-shoe structure, are used to support a Ka-band corrugated feed horn. The horse-shoe-shaped main tube structure and the asymmetrically connected tripod tubes are uniquely designed in geometry to avoid membrane damage and flatness deviation when the deflated antenna structure is rolled up.

The above Ka-band 3-m inflatable reflectarray was built for primarily laboratory demonstration of its RF performance only. Since then, a second model

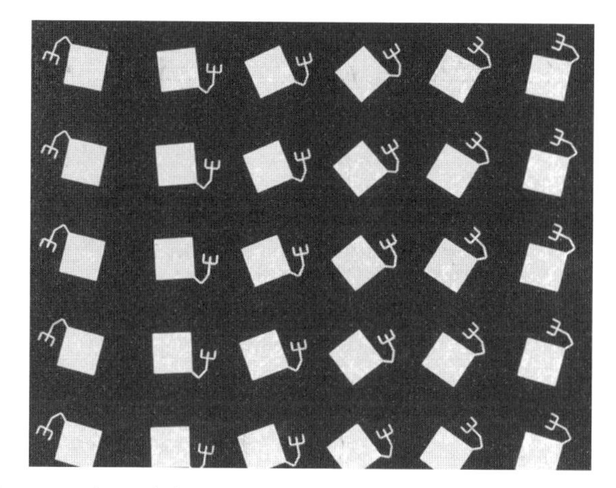

Figure 7.4. Close-up view of the Ka-band CP patch elements having different angular rotations.

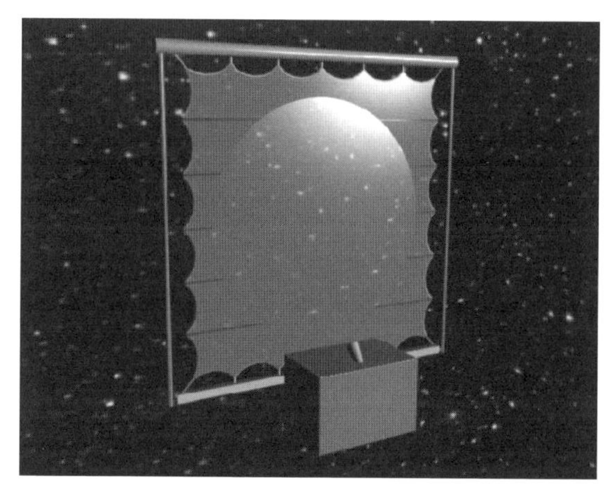

Figure 7.5. Conceptual drawing of inflatable reflectarray showing the feed horn being fixed on the spacecraft.

[5] was developed to demonstrate its mechanical integrity for space application. There are two major differences. One is that the second model has its inflatable reflectarray surface deployed without the deployment of a tripod-supported feed. The offset feed is fixed on the spacecraft bus, as illustrated in Fig. 7.5, where the inflatable booms and reflectarray membrane surface can be rolled up and down as a movie screen. The second major difference is that the inflatable tubes are made of rigidizable aluminum reinforced internally by using carpenter tapes, as shown in Fig. 7.6. This type of tube is called a spring-

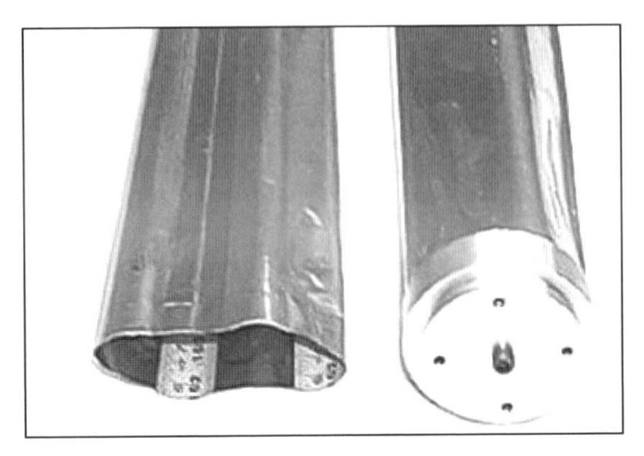

Figure 7.6 Inflatable boom showing rigidizable aluminum membrane and carpenter tapes for reinforcement.

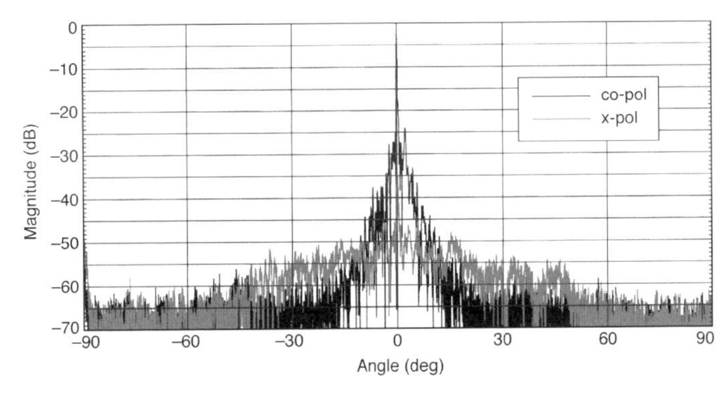

Figure 7.7. Measured far-field pattern of the 3-m inflatable reflectarray with circular polarization.

tape reinforced (STR) boom [6]. Once the booms are inflated in space, the aluminum membrane will soon rigidize itself and the inflation gas is no longer needed. In addition, in the event that the tubes are penetrated by small space debris, they will remain rigid to provide proper support for the reflectarray membranes. The carpenter tapes are used as reinforcement to provide additional axial load capacity as well as some orthogonal load capacity to each tube.

The antenna's RF tests were performed at an indoor compact range where up to 10-m size antennas can be tested. A typical elevation pattern of the antenna is given in Fig. 7.7 where a 0.22° beamwidth was measured. The sidelobe level is −25 dB or lower below the main beam peak, and the cross-polarization level is −40 dB or lower. The antenna gain was measured versus frequency. The results show that the antenna is tuned to the desired frequency of 32.0 GHz with a –3 dB bandwidth of 550 MHz. A peak gain of 54.4 dBic was measured. This measured antenna gain indicates an aperture efficiency of 30 percent, which is lower than the expected 40 percent. This relatively lower efficiency was the result of large element resistive loss due to the poor loss-tangent (≈0.03) of Kapton substrate material used. A recent commercial available thin-membrane material called Liquid Crystal Polymer (LCP) with a loss tangent of 0.002 is a better choice for the future development. It is quite certain that future development can improve the efficiency to the expected 40 percent or higher. The measured surface flatness data of the antenna aperture shows a root-mean-square (RMS) value of 0.2 mm, while the required surface RMS value is 0.5 mm. This good surface flatness is also reflected by the well-formed far-field pattern with expected main beamwidth and low sidelobe level. Although the aperture efficiency was not as expected, the achievement of excellent membrane flatness indicates that an inflatable array antenna at Ka-band is now feasible.

7.2 CONTOURED BEAM REFLECTARRAYS FOR SPACE APPLICATIONS

Contoured beams can be easily achieved using reflectarrays by implementing an appropriate phase-shift on the reflectarray elements. In reflectarrays, as opposed to phased arrays, only the phase at each element can be adjusted with the amplitude determined by the feed horn. On the other hand, the reflectarray is made up of many phase-shifter elements, which provide many degrees of freedom for shaping the beam. Contoured beam reflectarray have been demonstrated for a cosecant squared pattern using a 3600-element reflectarray in [7], and for Direct Broadcast Satellites (DBS) applications in [8].

For contoured-beam space applications such as DBS, currently shaped reflectors are satisfactorily used to provide coverage in a selected geographical region. However, as a drawback, a custom mold must be manufactured for the shaped reflector for each coverage specification, which increases both costs and manufacturing time. Reflectarrays, on the other hand, would eliminate the need of molds and allow reusing the mechanical models and tests for different missions, since only the dimensions of printed patches or lines are adjusted for a specified coverage and not the structural panel.

The contoured beam reflectarray reported in [8] was designed starting from a Ku-band shaped reflector previously manufactured for a DBS European coverage application. The required phase-shift at each reflectarray element was obtained from the distance between the shaped surface and a plane, where the reflectarray was placed to substitute for the reflector. A breadboard of axes 110 and 90 cm was designed at 14 GHz using a single layer of varying-sized patches, manufactured, and measured. A minimum gain of 23 dBi was achieved in the 99 percent of the coverage for a 7 percent bandwidth centered at 14.15 GHz. However, this reflectarray suffered from the bandwidth limitations mentioned in Chapter 5, and significant distortions of the pattern shape occur at different frequencies in the band. In addition, 7 dB gain variations were observed within the coverage regions. Two possible ways of improvement were suggested in [8], first to use a direct array synthesis method to obtain the phase distribution on the reflectarray without the previous design of a shaped reflector; and, second, to perform optimizations in a frequency band by taking into account the frequency dependence of reflectarray elements, in a similar way as the shape of a reflector is optimized for a given frequency band. The second approach was already discussed in Chapter 5 and requires a multilayer configuration to provide more degrees of freedom, while the first one is addressed here.

A direct synthesis method for shaped-beam reflectarrays will provide some advantages. First, the previous design of an equivalent shaped reflector is not required. Second, small errors in computing the phase on the reflectarray by using distances are eliminated. Third, the phase synthesis is not constrained by geometrical parameters and therefore is more flexible for synthesizing any required radiation pattern. Finally, the phase for both linear polarizations is synthesized independently and, hence, a different pattern for each polarization can be achieved, as demonstrated in [9]. Also, possible differences between the two linear polarizations in the radiation pattern or phase center of the horn can be taken into account during the reflectarray design. In addition, the reflectarray can be designed to change the polarization, for example, to convert linear into circular polarization just by adding a 90° phase-shift in one linear polarization with respect to the orthogonal one.

In reflectarrays, the synthesis of radiation patterns is restricted by the feed since it fixes the amplitude of the incident field on each reflectarray element, and only the phase distribution can be modified. To implement the required phase-shift distribution, the dimensions of the corresponding geometrical parameter must be adjusted in the reflectarray elements. A direct optimization process, in which the element dimensions are adjusted in an iterative process to achieve the required contoured pattern, is computationally unaffordable, because it would require the analysis of all the elements (several thousands in space applications) to compute the radiation pattern in each iteration of the pattern synthesis.

An efficient alternative consists of dividing the design process into two steps as described in [10]. In the first step, assuming a fixed amplitude distribution on the reflectarray surface given by the feed radiation pattern, a phase-only synthesis is applied to compute the phase of the reflected field at each reflectarray element that provides the required contoured pattern. In the second step, the patch dimensions are optimized element by element to achieve the previous phase distribution and the required frequency variations in a given bandwidth, following the same process as the one described in Chapter 5 for a pencil-beam reflectarray.

7.2.1 Pattern Synthesis

For the pattern synthesis in reflectarrays, the amplitude of the incident field on the individual cells is fixed by the radiation pattern of the feed, and the amplitude of the reflection coefficient in each cell is practically unity, except for a small reduction produced by the element losses. Thus, the only variables to be optimized are the phases of the reflection coefficients, or phase-shift, at the reflectarray elements, then a *phase-only* synthesis must be applied. Several phase-only techniques have been developed to obtain shaped beams with planar arrays [11–13].

The same techniques can be applied to obtain the required phase-shift distribution in reflectarrays, but the problem is more challenging because of the very high number of elements in reflectarrays, particularly for space applications. Therefore, only very efficient techniques able to deal with thousands of variables can be applied to reflectarray's pattern-synthesis. It has been demonstrated that the phase-only technique known as the *Intersection Approach* [14], previously developed for phased array, is very efficient for large reflectarrays [15, 16] and it is briefly described here.

First, the contoured beam requirements defined in the satellite or antenna coordinate system are transformed to the reflectarray coordinate system. The Satellite (S) coordinate system has the Z_S-axis pointing towards the center of the Earth and the X_S axis pointing towards the West in the equatorial plane, the Antenna (A) coordinate system is defined by pointing the Z_A-axis to a central point in the required coverage and by a rotation around the Z_A-axis (Fig. 7.8). The Reflectarray (R) coordinate system is defined with X_R and Y_R axes on the reflectarray surface, as shown in Fig. 7.9. As an example, see the coverage in S coordinate system shown in Fig. 7.10 for South America and Florida seen from the 67° West orbital position. This coverage includes three

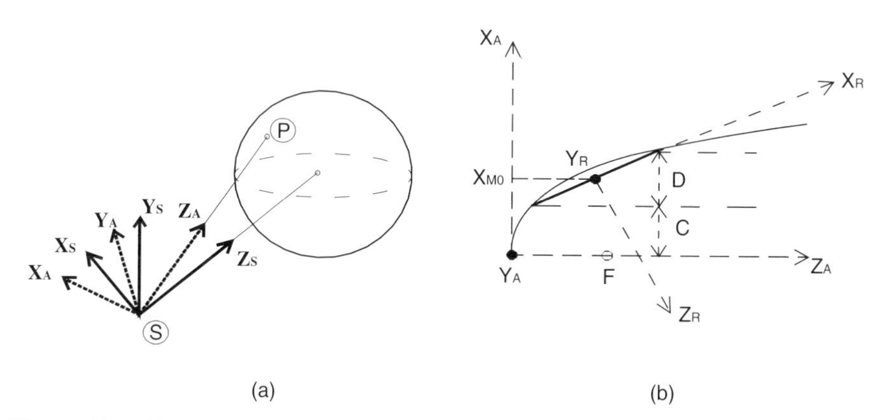

Figure 7.8. Different coordinate systems. (a) Satellite and antenna coordinate systems, (b) Antenna and reflectarray coordinate systems.

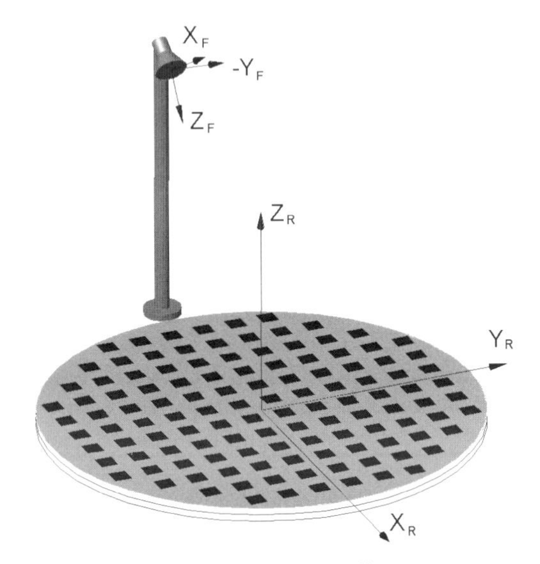

Figure 7.9. Reflectarray coordinate system.

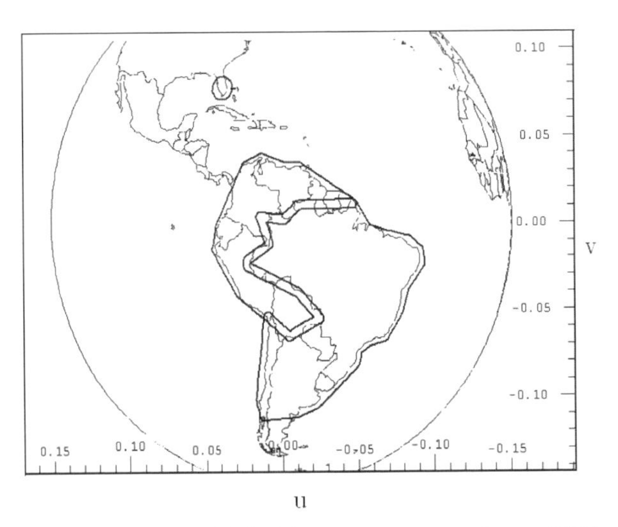

Figure 7.10. South America and Florida coverage seen from 67°W geostationary orbital position.

contours, shown in the figure with thicker lines, with a minimum gain of 25 dBi for the Eastern continental region and 22 dBi for the other two regions.

The contour requirements in the R coordinate system, are specified by a *mask* with minimum and maximum values of gain in a region of the u-v plane. Fig. 7.11 shows the mask for minimum gain in the coverage regions; the

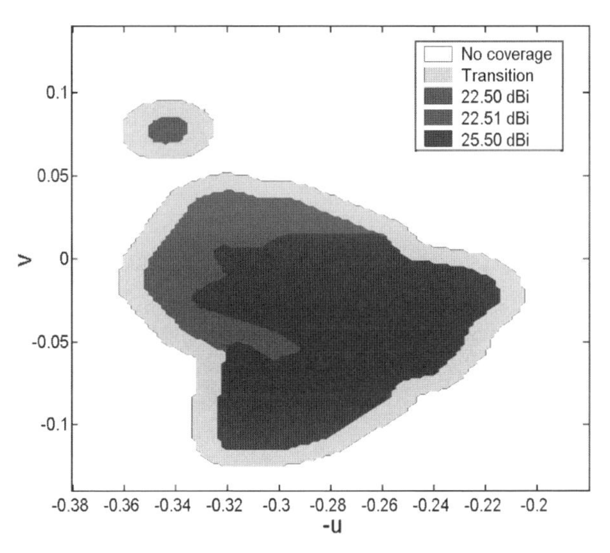

Figure 7.11. Gain requirements in the reflectarray coordinate system.

maximum gain is fixed to 1 dB higher. A transition region is defined around the coverage regions with no gain requirements, to allow an appropriate roll-off to the side-lobe region.

In the *Intersection Approach* technique, two sets are considered: the set of the radiation patterns that can be obtained with the reflectarray (R_S), and the set of the radiation patterns that fulfill the required mask of specifications (M_S). The synthesis consists of finding an element in set R_S that also belongs to M_S, or within a minimum distance from set M_S, by alternative projecting from R_S to M_S and vice versa. To accomplish that task with flexibility (introduction of constraints in the amplitude imposed the feed) and efficiency (fast algorithm even for large number of elements), suitable definitions of the corresponding projector operators are required, as described in [14, 15].

The choice of the starting pattern belonging to set R_S becomes very important in order to avoid that the iterative process converges to nondesired local minima. Two different options have been considered. The first one takes as the starting phase distribution, that one corresponding to the superposition of the aperture fields associated with several pencil beams very close to each other that approximates the desired shaped beam [17]. This initial phase distribution already provides a shaped beam but with high side lobes. The second option, simpler than the first, considers an out-of-focus beam properly directed. The results have demonstrated that both options for the starting point lead to good final patterns. The initial radiation pattern is projected into set M_S, by modifying the gain up to the *mask* levels. From this modified radiation pattern that fulfills the *mask*, both amplitude and phase distributions are obtained by using an FFT algorithm. The projection into the R_S set is performed by changing the amplitude at each element to the value imposed by the feed. The process is repeated until a minimum distance between the two sets is reached.

In many cases, the method converges into a local minimum, which does not satisfy the mask requirements. The number of local minima rapidly increases with the number of variables in the optimization, that is, the number of reflectarray elements. When applying the synthesis technique to a reflectarray with thousands of elements, it is very common to end in a local minimum with narrow hollows inside the shaped beam [18]. To overcome this inconvenience, a process in several stages where the illumination on the reflectarray's edge is gradually increased from one step to the next was proposed in [10] and [15]. In the first stage, a very high taper is chosen for the reflectarray illumination, so that the amplitude levels near the reflectarray's border are very low, and hence only the central part of the reflectarray contributes significantly to the radiation patterns. This is equivalent to reducing the size of the reflectarray, since the outer elements are excited with negligible amplitude, and then the number of local minima is reduced. The taper for the first step is chosen so that no holes appear in the coverage region after applying the *Intersection Approach*. In the following stages, the taper level at the border is gradually increased and the phase distribution obtained in the previous stage is used as the starting point for the *Intersection Approach*. The procedure finishes with the phase-only synthesis for the specified taper corresponding to the feed-horn.

The radiation pattern of the feed is typically modeled as a cosine function to the exponent q, which facilitates the variation of the amplitude's taper. The higher the taper, that is, the q factor, the smaller the reflectarray's edge illumination. The flow chart associated with the stepped pattern synthesis is shown in Fig. 7.12. Although convergence to a global minimum is not ensured by this technique, it has been checked with several examples that the problem of local minima in large reflectarrays is significantly reduced by this multistep process [9, 10, 15, 19, 20] and the method yields a phase distribution that provides a radiation pattern that fulfils the mask of requirements. For a more accurate modeling of the horn, a final stage of the stepped synthesis process can be realized where the cosine model is substituted by the real field pattern, obtained from simulations or measurements. This last stage does not lead to significant variations of the synthesized phase distribution if the radiation pattern of the feed-horn exhibits quasi axis-symmetry, as is the case for corrugated circular horns.

In the previously described phase-only synthesis method, the synthesized phase distribution at a generic k-step, $\phi_K(i, j)$, depends on the phases of the preceding iteration $\phi_{K-1}(i, j)$, and it can experiment with changes of 360° in any reflectarray element (i, j) with respect to that in the previous step. Actually, the possible variations oscillate from –180° to 180° due to the *arc tan* function used in the implementation of the alternating projection algorithm. These large variations in phase from one iteration to the next can cause convergence

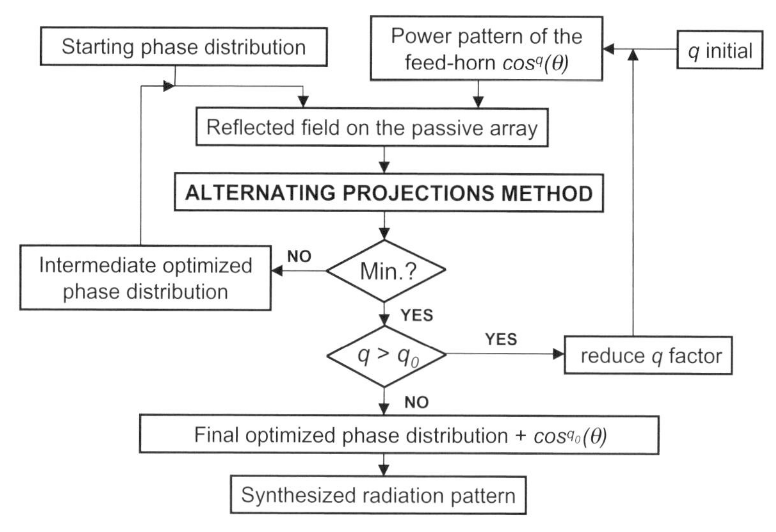

Figure 7.12. Scheme of phase-only synthesis technique in several stages.

problems. However, phase constraints can be implemented in the process by limiting the range of phase variations to a smaller range [21]. This option is particularly interesting when the starting phase distribution in a particular synthesis stage is already close to provide a pattern fulfilling the contour requirements, and therefore small variations in the phase distribution are sufficient to fulfill the requirements. In addition, this limitation in the phase variations can be used to obtain a phase distribution with a smooth variation on the reflectarray surface after the pattern synthesis, which is more convenient for the design of the reflective elements, particularly if a local periodicity approach is used.

To implement the limits in phase variation, the difference between initial phase $\phi_1(i, j)$ and phase at K-step $\phi_K(i, j)$ is first computed and limited to the range (−180°, 180), then, during the projection on the set of possible reflectarray patterns (R_S), the phase is enforced to verify in each iteration the following condition

$$-M < \phi_1(i, j) - \phi_k(i, j) < M, \quad k \neq 1, \tag{7.1}$$

where M is a positive real magnitude less than 180°. For the particular case M = 180, the phase-only synthesis would not be phase-constrained. The difference of the bounded synthesized phase distribution in each iteration $\phi_K(i, j)$ is always referred to the starting phase $\phi_1(i, j)$, since the purpose of the synthesis method is to achieve a final phase distribution with an absolute phase variation less than M degrees with respect to the initial phase used in the synthesis. The phase constraint in the alternating projection algorithm does not introduce an increase in the computational cost, and the reflectarray synthesis is highly efficient even for very large reflectarrays. Two particular cases of interest can take advantage of this synthesis with phase constraints: first, reconfigurable reflectarrays using controllable phase shifters in a limited range of phase variations; and second, multifrequency pattern synthesis, which is described below.

For telecommunications antennas in space, the contoured requirements must be fulfilled in a given frequency band, or in two separated bands for transmit/receive antennas. When reflectarrays are designed for these applications, first the required phase-shift must be found at several frequencies in the working band, and then the reflectarray elements must be optimized to provide those phase distributions. For the problem of pattern synthesis at several frequencies, first the *Intersection Approach* is applied at the central frequency, and then, assuming that the phase delay varies with frequency as in the case of a pencil beam, that is, the variation provided by an ideal parabolic reflector, the required phase shift on each reflectarray element is computed at any frequency in the working band [22]. Using this approach, the contours are well matched and beam squint is eliminated in a 10 percent bandwidth [10]. However for frequencies farther away from the central frequency, the contours are distorted. This limitation is overcome by using the previous phase distribution as a starting point for the *Intersection Approach* algorithm, which is applied again at each frequency f_i. Since the initial phase distribution at frequency f_i is close to the final solution, the *Intersection Approach* is applied in only one step, using the real illumination taper and limiting the maximum variation in phase. In a few iterations, the algorithm rapidly converges to a solution that fulfill the requirements. The multifrequency pattern synthesis has been applied to obtain the phase distribution at frequencies of 10.7 and 13.5 GHz for a given European coverage [21, 23]. First, the phase distribution was obtained at the central frequency (12.1 GHz), and then at extreme frequencies. The proposed process ensures a smooth variation of phase distribution with frequency, which is important to the design of the reflectarray patches. This technique can be applied to both broadband and multifrequency applications.

7.2.2 Practical Designs

The above outlined procedure has been applied to designing a number of shaped beam reflectarrays for DBS applications, which are presented here. In all cases, the reflectarray is made of three stacked layers of printed arrays with rectangular patches, in which the dimensions of the staked patches are optimized in each element to ensure the required coverage at several frequencies in the working band. The periodic cell is shown in Fig. 7.13. For the first example, described in [10], the pattern requirements are shown in Fig. 7.10. An elliptical reflectarray with axes 83.2 cm × 80.6 cm has been designed for dual polarization in the 12.8–14.2 GHz band (10 percent). The feed is located

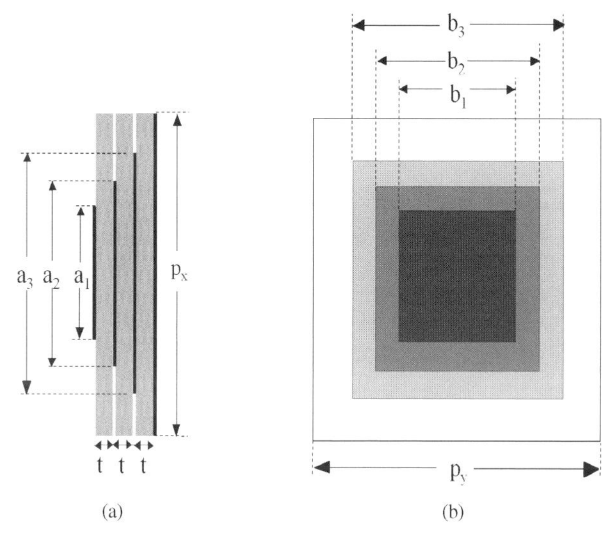

Figure 7.13. Periodic cell of three-layer reflectarray with varying-sized patches. (a) Lateral view, (b) Front view.

at coordinates $x_f = -32.45$ cm, $y_f = 0$ cm, $z_f = 100.56$ cm, and its radiation pattern is simulated as a $cos^q(\theta)$, with a −15 dB taper at the reflectarray border. The patches in each layer are in a 13 mm × 13 mm square lattice, printed on a Kapton film with 0.1-mm Kevlar skin on both sides, and separated by 2-mm-thick Kevlar honeycomb with dielectric constant $\varepsilon_r = 1.1$ and loss tangent equal to 0.002. Starting from an out-of-focus pencil-beam phase distribution, the required phase-shift distribution shown in Fig. 7.14 is obtained after applying the synthesis procedure at 13.5 GHz.

These phases are valid for both linear polarizations, assuming the same radiation pattern of the horn for the two polarizations. First, the patch dimensions are adjusted to produce the phase-shift distribution shown in Fig. 7.14 at the central frequency, but maintaining a fixed relative patch size in each stacked array, ($a_1 = 0.6a_3$, $a_2 = 0.9a_3$, $b_1 = 0.6b_3$, $b_2 = 0.9b_3$, being $a_i \times b_i$ the dimensions of a patch at layer i). The analysis routine is based on the Spectral-Domain Method of Moments assuming local periodicity. The radiation pattern contours (25, 22, and 8 dBi) for X-polarization (with the electric field in x-direction) are represented in Fig. 7.15 at 12.8, 13.5, and 14.2 GHz together with the coverage requirements as a function of normalized angular coordinates (u, v) referred to the reflectarray coordinate system. The magnitude represented here is the gain of the antenna in dBi computed as described in Section 3.8. The pattern at 13.5 GHz fulfills the gain values given in Fig. 7.11, but due to the narrow band behavior, the patterns are completely distorted at the two extremes of the band.

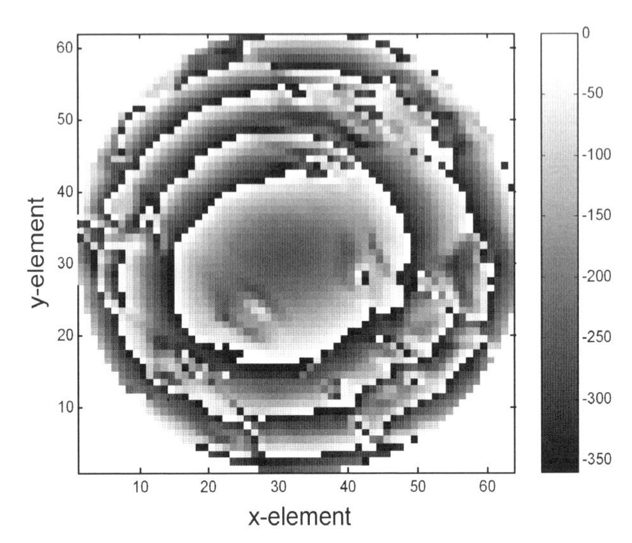

Figure 7.14. Required phase-shift distribution at 13.5 GHz to generate the contoured beam for the South America coverage.

Assuming an ideal variation of the phase with the frequency, which is the one corresponding to a parabolic reflector, the required phase distributions are computed at extreme frequencies, and it was checked that the radiation patterns fit very well the mask contours, but at the higher frequency the overall radiation pattern is slightly narrower because the antenna dimensions are larger in wavelengths. The optimization technique described in Chapter 5 is applied and the dimensions of the patches in the three layers are adjusted independently to match the required phase-shift at 13.5 GHz and the ideal difference in phase delay at the extreme frequencies, for both polarizations.

After the optimization of patch dimensions, the errors in the phase, defined as the difference between the required phase and the one obtained by MoM, are drastically reduced at the extreme frequencies. The resulting radiation patterns are shown in Fig. 7.16 for X-polarization at central and extreme frequencies. Similar radiation patterns are obtained for the orthogonal polarization. The photo-etching mask of the first array layer is shown in Fig. 7.17. When comparing the patterns of Fig. 7.16 with those of Fig. 7.15, the improvement obtained in the optimization process is clear. Now, the reflectarray practically fulfills the gain requirements in the working frequency band (10 percent), but with small distortions at the extreme frequencies.

A further improvement can be achieved by matching several frequencies in the working band instead of only the difference in phase delay at extreme frequencies. The cross-polarization levels are always −30 dB below the maximum gain. A further improvement in cross-polar isolation can be achieved by using separate feeds for the two polarizations.

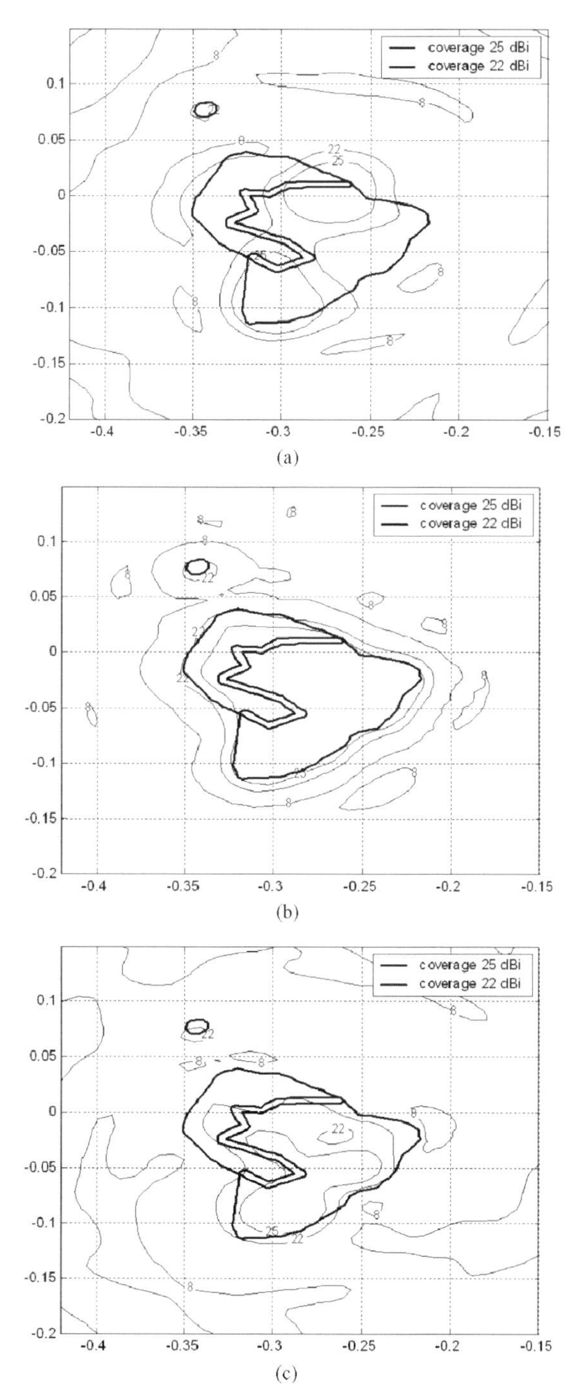

Figure 7.15. Gain radiation patterns at central and extreme frequencies for the reflectarray designed at 13.5 GHz. (a) 12.8 GHz, (b) 13.5 GHz, (c) 14.2 GHz.

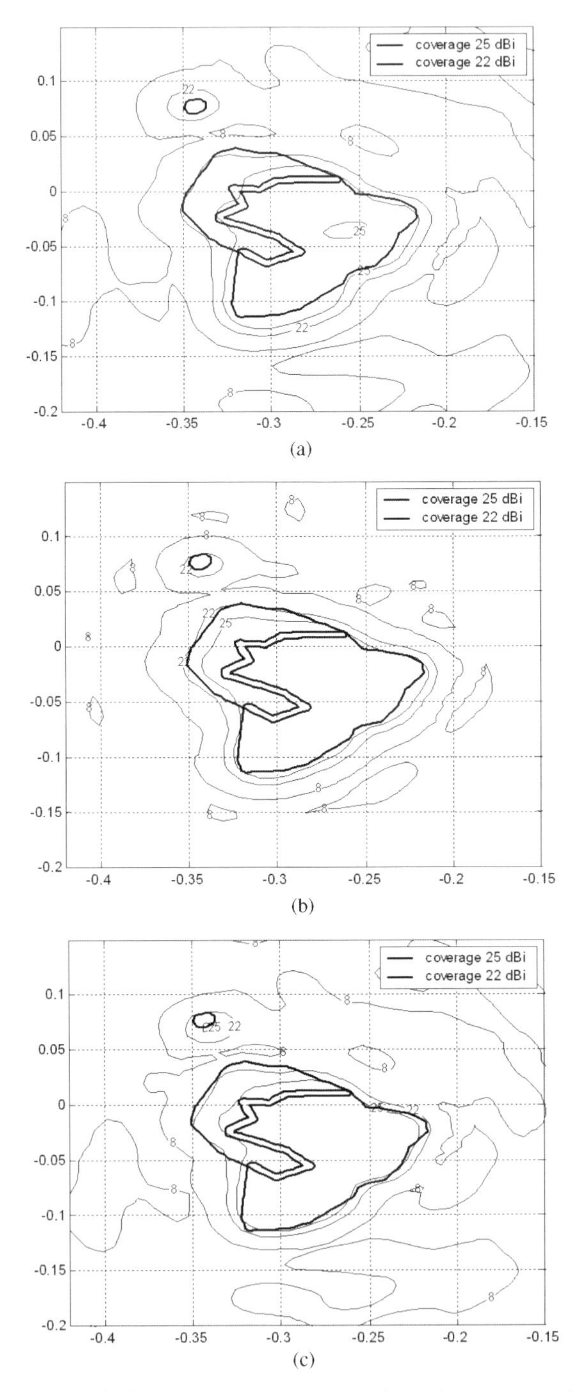

Figure 7.16. Gain radiation patterns at central and extreme frequencies for the reflectarray optimized in the 12.8–14.2 GHz band. (a) 12.8 GHz, (b) 13.5 GHz, (c) 14.2 GHz.

Figure 7.17. Photo-etching mask for the first layer of the optimized reflectarray.

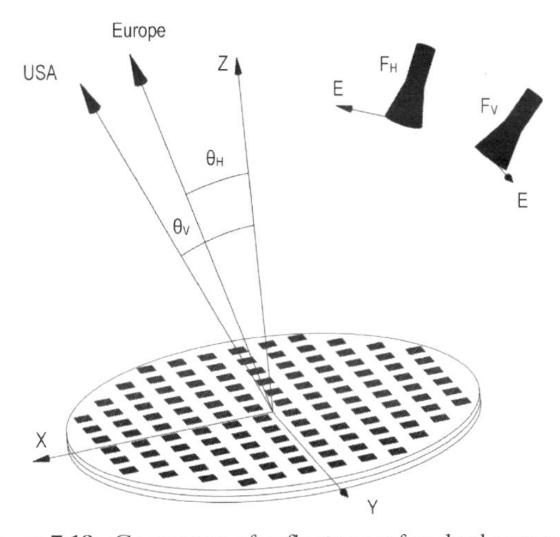

Figure 7.18. Geometry of reflectarray for dual coverage.

The second example is a breadboard of a DBS transmit antenna with different coverages in each polarization. The breadboard has been designed, manufactured using Space proven technology and tested, and the results compared with a conventional dual-gridded reflector [20]. In this case, the reflectarray is dual-fed, that is, each coverage is obtained by using an independent feed horn as shown in Fig. 7.18. The coverage requirements for horizontal (H-) and vertical (V-) polarization are shown in Fig. 7.19 for a satellite at 5° West orbital position. For H-polarization, the requirements consist of two gain

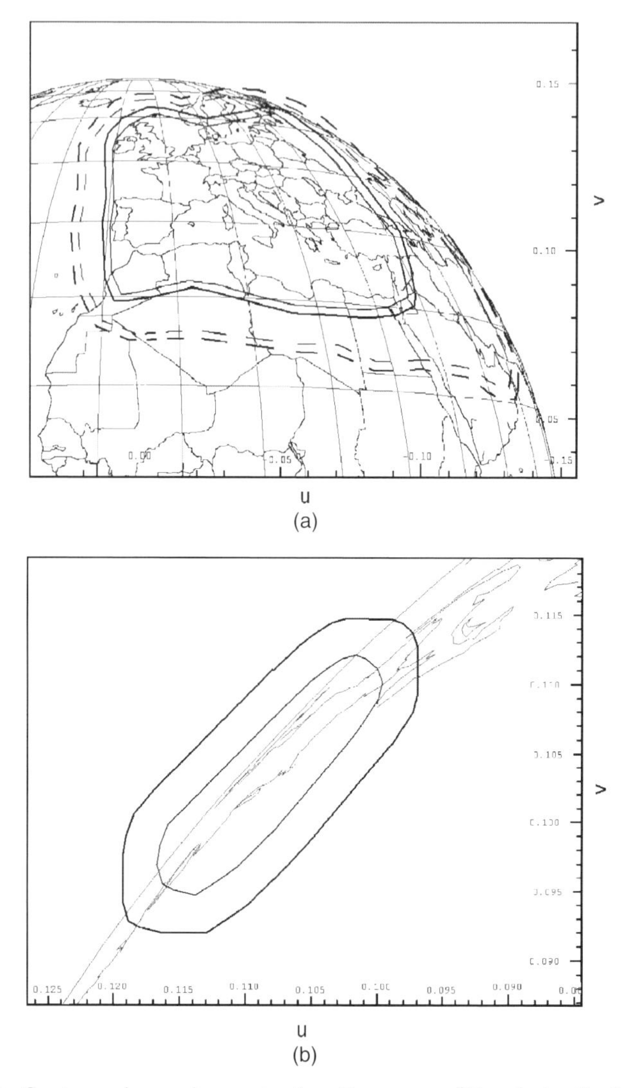

Figure 7.19. Contoured requirements for European (H-polarization) and North American (V-polarization) coverages.

contours of 28.5 dBi (solid line) and 25.5 dBi (dash line) in the 11.45–12.75 GHz frequency band, while 37 dBi are required in V-polarization in the North American coverage region (11.45–11.7 GHz). The outer contours represent the requirements after taking into account typical pointing errors of the satellite (0.1° in roll, 0.1° in pitch, and 0.5° in yaw). A cross-polar discrimination of 30 dB is required for both coverages.

In reflector technology, these requirements were achieved with a dual-gridded reflector with a different reflecting surface for each polarization, which is used as reference. The novelty of this work is the demonstration that two independent beams, one for each polarization, can be generated with a single reflectarray surface and, consequently, a significant reduction in mass and volume is achieved.

The reflectarray consists of an elliptical flat panel with axes 1036 × 980 mm, which is the same aperture surface as in the dual-gridded antenna. The two feed horns are placed with the phase center on the XZ plane, at coordinates (in mm) $F_V = (-460, 0, 887)$ and $F_H = (-302, 0, 898)$ for V- and H-polarization, respectively, so that the projection of the field radiated by each feed on the XY plane is parallel to one side of the rectangular patches, and then the phase shift for each polarization can be controlled independently by each patch dimension. The period for the reflective elements is defined as 14 × 14 mm (0.6λ at 12.75 GHz and 0.53λ at 11.45 GHz). Both feed locations and period have been chosen to avoid the appearance of grating lobes at any reflectarray element, considering the angle of incidence of the field coming from the feed.

For the contoured European beam, the *Intersection Approach* has been applied at the central frequency to obtain the phase distribution on the reflectarray surface shown in Fig. 7.20(a), while for V-polarization, the phase distribution shown in Fig. 7.20(b), which corresponds to a pencil beam, is considered because it provides the maximum gain and fulfills the gain requirements. The required phases at extreme frequencies were computed by multifrequency synthesis for H-polarization, and by assuming the ideal linear variation for the pencil beam in V-polarization.

The reflectarray panel is made of an electrical sandwich with three array layers over a ground plane and backside stiffening sandwich with carbon fiber-reinforced plastic (CFRP), as shown Fig. 7.21. The electrical layers were defined using Kapton with copper patches and a Kevlar/resin stiffening layer on one side only. To reduce dissipative losses, a low-loss resin (cyanate-ester) is chosen for the Kevlar composite layers and also for bonding the array layers to Nomex honeycomb 2-mm thick, used as separators. The CFRP face-sheet (layer A in Fig. 7.21) was used as the ground plane, with an increase of dissipative losses around 0.1 dB, which is a typical value in CFRP reflectors. The reflectarray was designed by optimizing the dimensions of the rectangular patches to achieve the required phase shift at the central and extreme frequencies for each polarization (11.45–12.75 GHz for H-polarization and 11.05–12.1 for V-polarization, which is larger than required), following the process described in [10] and [20]. The analysis technique used in the optimization process is a full-wave Method of Moments in spectral domain, assuming each element is in a periodic array environment, which takes into account the real angles of incidence at each element, the polarization of the incident field, and all the dielectric layers defined in the sandwich configuration.

After the optimization process, the array layers are manufactured by photo-etching, and the reflectarray panel was manufactured by a multistep curing

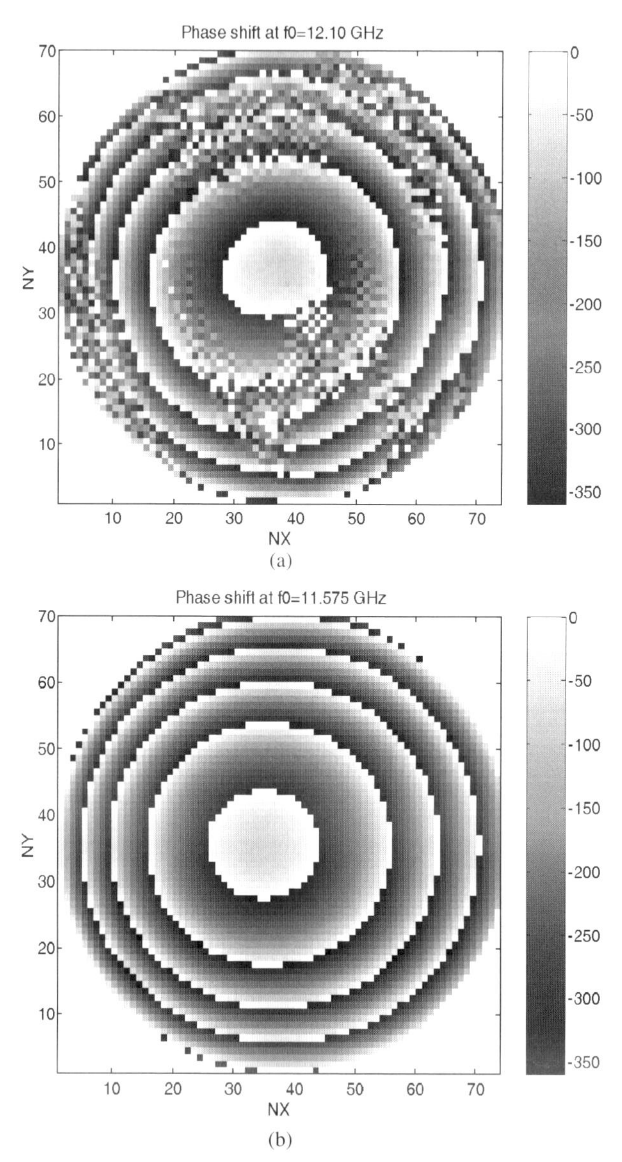

Figure 7.20 Required phase-shift at central frequency. (a) H-polarization (12.100 GHz), (b) V-polarization. (11.575 GHz).

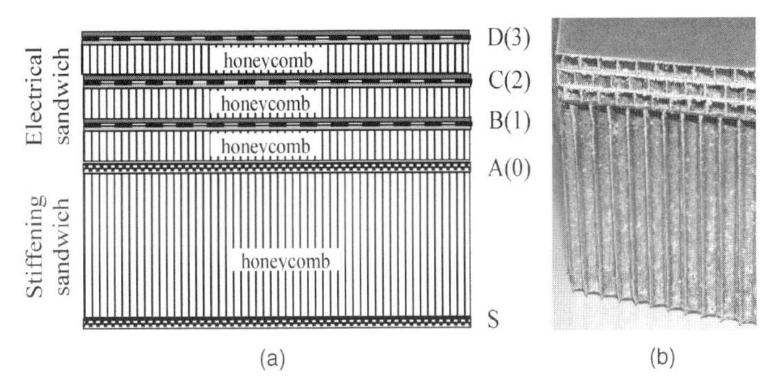

Figure 7.21. Configuration of reflectarray panel with electrical and stiffening layers. (a) Lay-up, (b) section of manufactured sandwiched.

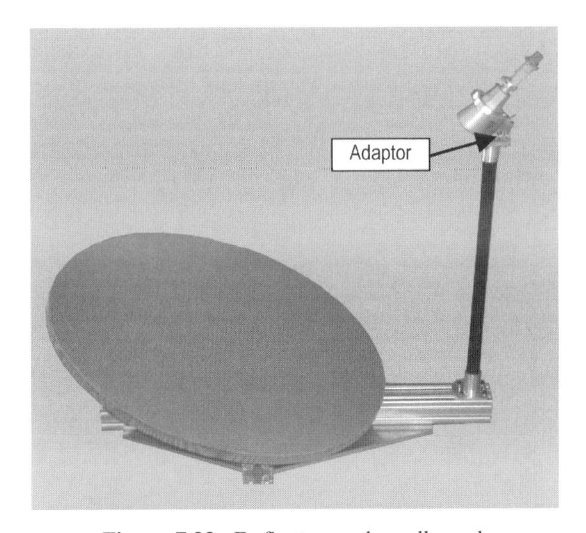

Figure 7.22. Reflectarray breadboard.

process in order to achieve maximum accuracy and repeatability in the thickness and composition of each electrical layer. The breadboard was assembled and measurements were made in H- and V-polarization using the same feed in different positions defined by the a displacement of the arm and by changing the adaptor used to fix the feed (Fig. 7.22).

The measured co-polar contour gain patterns at extreme frequencies are shown in Fig. 7.23 for V-polarization. The gain requirements of 37 dBi are fulfilled in the whole frequency band used in the design process (11.05–12.1 GHz). The cross-polarization levels in the U.S. coverage are always lower than 3 dBi in the whole frequency band, which corresponds to a cross-polar isolation

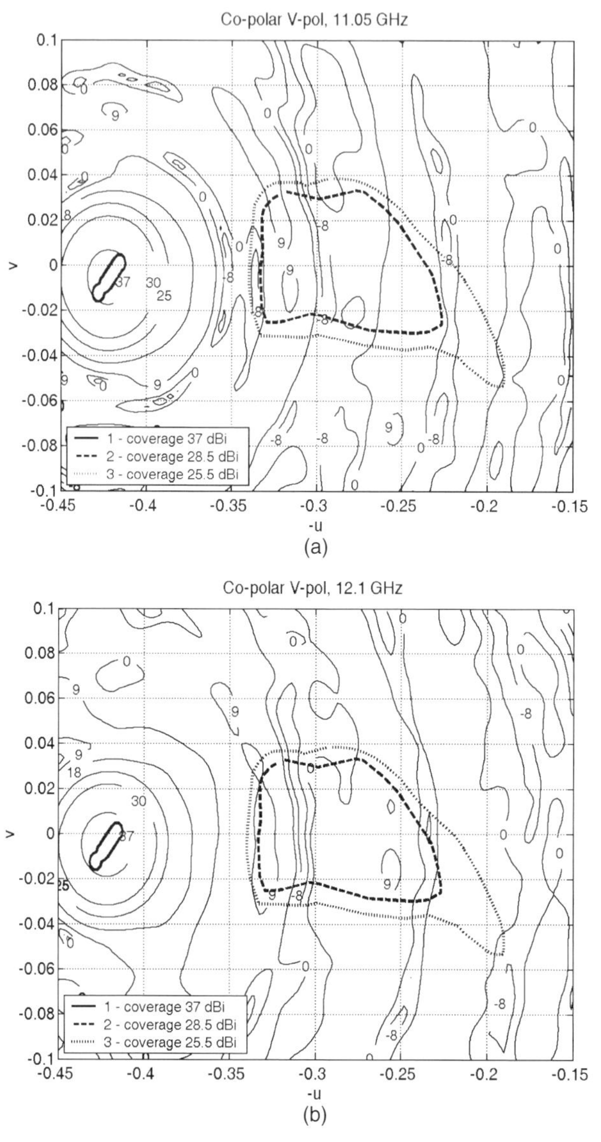

Figure 7.23. Measured co-polar gain contour patterns for V-polarization at extreme frequencies. (a) At 11.05 GHz, (b) at 12.1 GHz.

better than 34 dB. The cross-polar patterns are shown in Fig. 7.24 for the central frequency.

The measured co-polar patterns for H-polarization is compared with the simulations for three gain levels (28, 25, and 20 dBi) at 12.1 GHz (Fig. 7.25).

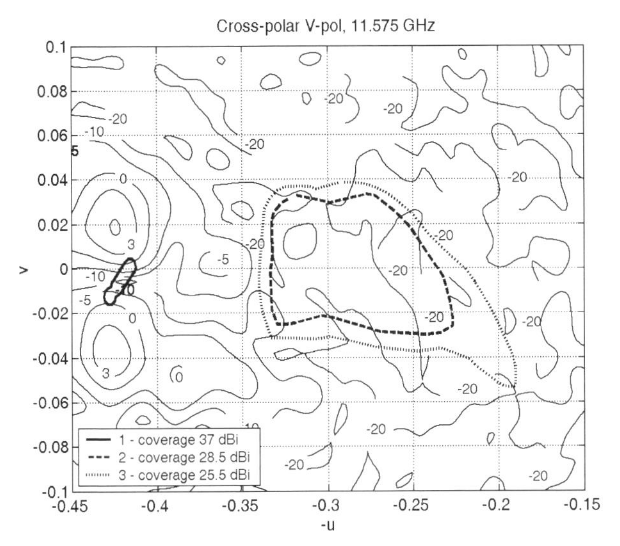

Figure 7.24. Measured cross-polar patterns in dBi for V-polarization at central frequency (11.575 GHz).

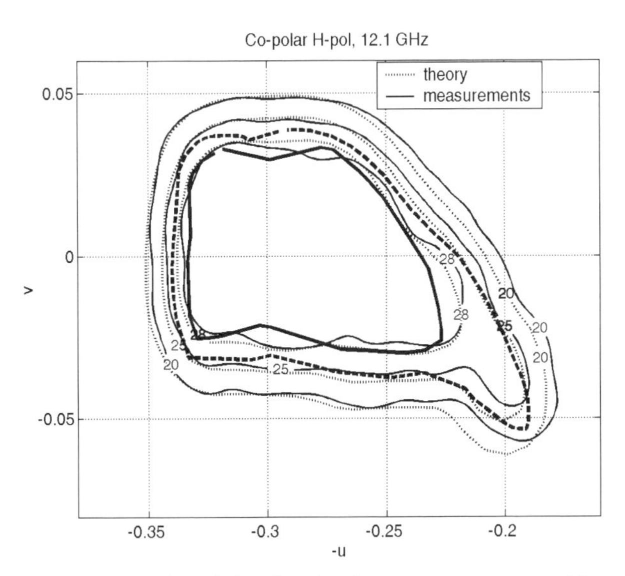

Figure 7.25. Measured and simulated gain contours at central frequency.

The gain patterns practically fulfill the mask of requirements in close agreement with the simulated radiation patterns using the analysis routine based on Method of Moments. The contour of 28.5 dBi is, however obtained with 28 dBi of gain in 99 percent of the enlarged coverage. The reduction of 0.5 dB in gain is the result of dissipative losses (0.4 dB) and the small errors in phase after the optimizations of the patch dimensions. The gain patterns practically fulfill the mask of requirements from 11.7 to 13 GHz, as shown in Fig. 7.26. However, a small shift in the frequency band to higher frequencies is observed, because the breadboard was designed in the 11.45–12.75 GHz band. It must be noticed that the frequency shift was observed only in H-polarization, and the reason for that must be the anisotropy of the honeycomb produced by its hexagonal structure, which means that the dielectric constant is higher with the electric field along the ribbon, with typical values from $\varepsilon_r = 1.06$ to $\varepsilon_r = 1.1$. Also, the resin used for bonding to the honeycomb is distributed along the hexagonal cells and will also exhibit a similar anisotropy.

In the breadboard, the electric field is across the ribbon for H-polarization and along the ribbon for V-polarization. Since honeycomb and resin anisotropy was not considered in the design and $\varepsilon_r = 1.1$ was assumed for both polarizations, the agreement is good for V-polarization, while a frequency shift is produced in H-polarization as a result of the lower ε_r. In addition, the small error in the ε_r for H-polarization can also be the reason for the small deviations in the contour patterns, as a result of the nonlinear behavior of the phase.

The radiation patterns were also measured at 12.1 GHz for H-polarization in the whole angular range in a spherical near-field system, in order to check possible spurious radiation out of the coverage region, and to measure the directivity and the dissipative losses by difference with measured gain. Measured losses are 0.35 dB with an accuracy in the measurements of ±0.18 dB, which is in close agreement with the predicted value of 0.4 dB. Fig. 7.27 shows the cuts of the co- and cross-polar radiation patterns at the different phi planes in the whole angular range ($0° \leq \theta \leq 90°$, $0° \leq \varphi \leq 360°$). This figure shows that there are no spurious lobes out of the coverage region, and also that the maximum levels of cross-polarization are better than −28 dB below the maximum gain.

This breadboard has demonstrated the ability of reflectarrays to realize two linearly polarized independent beams with satisfactory cross-polarization levels. The electrical performance has been compared with those of the reference dual-gridded antenna. For the directive beam in V-polarization, the gain measured in the North American coverage is comparable with that measured in the dual-gridded reflector. For the contoured beam in H-polarization, the gain is slightly lower than for the dual-gridded reflector and small distortions are observed in the contour pattern, produced by the honeycomb anisotropy and manufacturing tolerances. The losses in the reflectarray (0.35 dB) are slightly higher than in the dual gridded reflector (0.10 dB for the front shell and 0.20 dB for the rear shell). The total mass of the reflectarray panel,

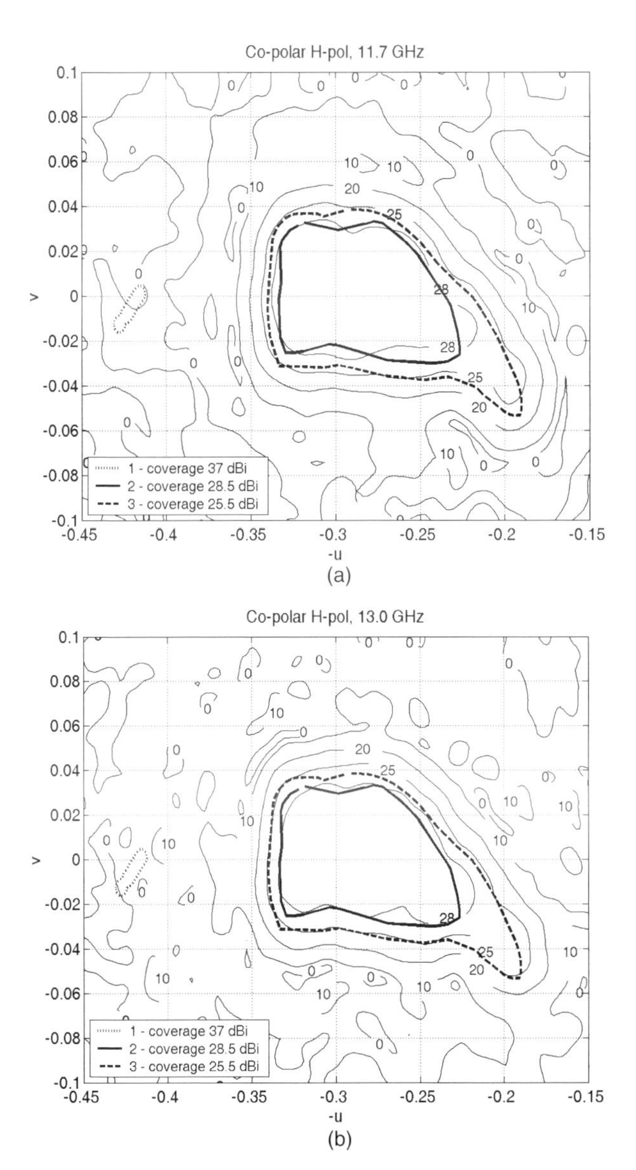

Figure 7.26. Measured co-polar gain patterns for H-polarization at extreme frequencies. (a) At 11.7 GHz, (b) at 13.0 GHz.

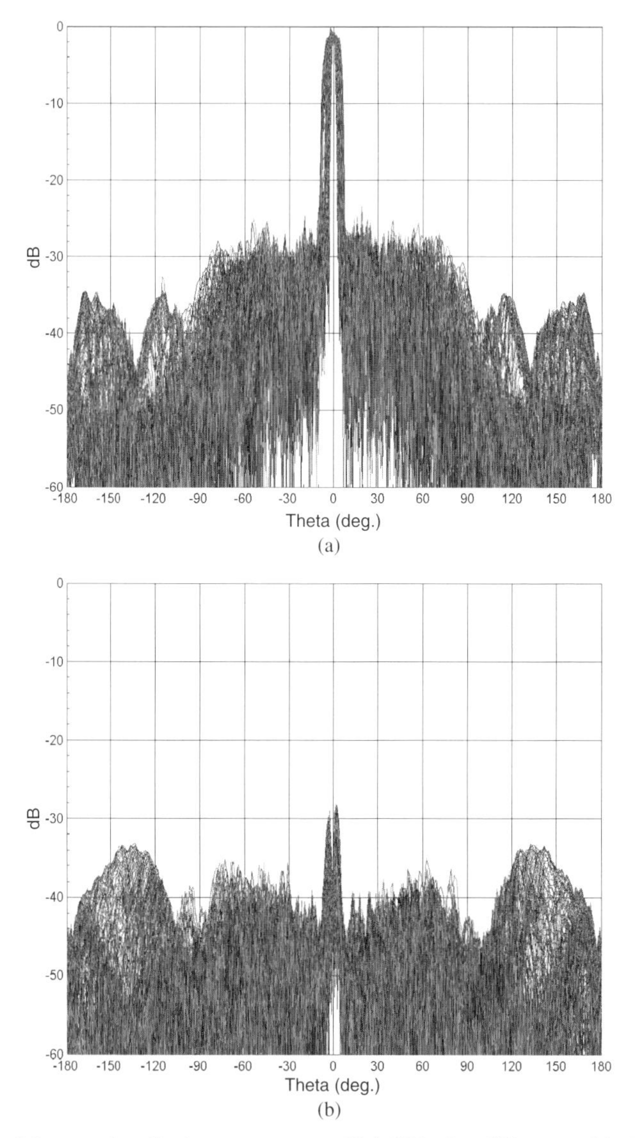

Figure 7.27. Measured radiation patterns at 12.1 GHz for different phi planes in the whole angular range ($0° \leq \theta \leq 90°$, $0° \leq \varphi \leq 360°$). (a) Co-polar patterns, (b) cross-polar patterns.

including electrical and structural sandwiches, is 2.250 kg, that is 2.7 kg/m^2, which represents a significant reduction in mass with respect to the dual-gridded antenna (4.3 kg/m^2 for the two-shell configuration).

The reflectarray breadboard has demonstrated the capacity to achieve satisfactory cross-polarization isolation, with cross-polar isolation better than −30 dB in the other coverage. For both polarizations, the performances remain stable over a 10 percent frequency band. Therefore, the reflectarray antenna can be an alternative to dual-gridded reflectors, characterized by a very low cross-polarization, but suffering from large volume and mass.

In the previous breadboard, a 10 percent bandwidth was achieved but it was demonstrated that larger bandwidth was not achievable by matching the phase at only central and extreme frequencies. In the next example, a 1-m reflectarray is designed for a DBS European coverage in a 15 percent frequency band by optimizing the patch dimensions to ensure the coverage requirements at five frequencies in the working band [24]. The coverage shown in Fig. 7.28 is referred to a satellite in the position: 10°E longitude, 0° latitude. The minimum gain requirements are 28 dBi on the coverage region defined by the gray line in Fig. 7.28, in the frequency band 10.95–12.75 GHz, and by the black line after taking into account typical satellite pointing errors. The antenna consists of a rectangular reflectarray panel (980 × 1036 mm) illuminated by a 20-dB Gaussian horn antenna from Flann Microwave with the phase-center positioned at coordinates $x_f = -358$, $y_f = 0$, $z_f = 1070$ mm. The feed was selected because its very low side-lobe levels and cross-polarization.

For the reflectarray design, the feed is modeled as a $\cos^q(\theta)$ function, where the q factor is determined at each frequency to match the measured patterns

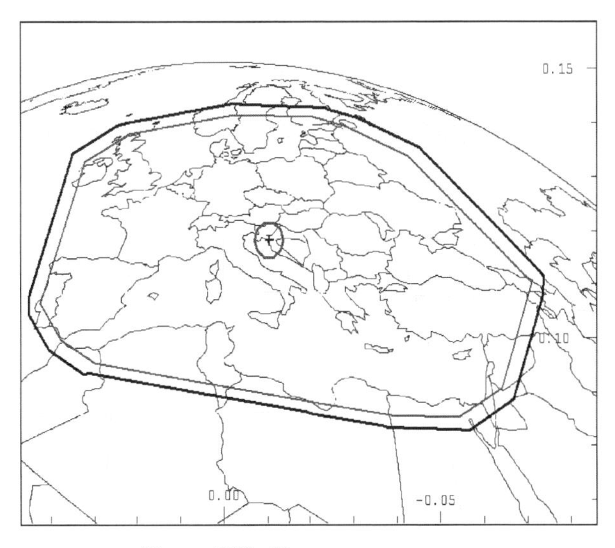

Figure 7.28. European coverage.

TABLE 7.1. Illumination level at reflectarray edges

Frequency (GHz)	10.95	11.4	11.85	12.3	12.75
q-Factor	18	21	23	28	32
Level at the edges (dB)	−14	−16.4	−17.9	−21.8	−24.9

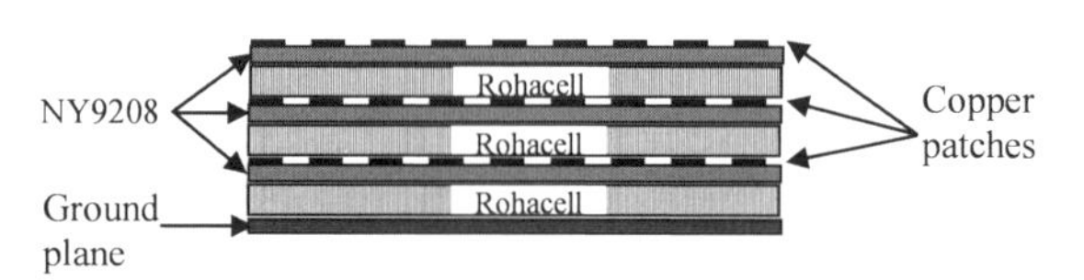

Figure 7.29. Sandwich configuration.

of the horn. The q factor and the illumination level at the edges of the reflectarray in the principal planes are shown in Table 7.1. The reflectarray panel is made of three stacked layers 0.762-mm-thick of a low-loss material (Neltec NY 9208, with $\varepsilon_r = 2.08$ and LT = 0.006) with the printed patches, separated by 2-mm-thick Rohacell HF31, as is shown in Fig. 7.29.

In order to achieve a 15 percent bandwidth, the dimensions of the staked patches are optimized in each element as described in Chapter 5 to match the required phase distribution at five frequencies (10.95, 11.4, 11.85, 12.3, and 12.75 GHz). To obtain the required phase distribution at those frequencies, multifrequency pattern synthesis [23] is applied. First, the *Intersection Approach* is applied at the central frequency (11.85 GHz), and then at 10.95 and 12.75 GHz, but limiting the phase variation to ±40°, as described in Section 7.2.1. For other frequencies in the 10.95–12.75 GHz band, the phase distribution has been obtained by linear interpolation from the phases at central and extreme frequencies, and the resulting radiation patterns fulfill the requirements in the whole frequency band.

After the optimization of the patch dimensions, the radiation patterns for H-polarization are shown in Fig. 7.30 at the central frequency and in Fig. 7.31 at the extreme frequencies. The radiation patterns are represented in gain, including the dissipative losses in the reflectarray, which is estimated as 0.13 dB. These results show that the gain requirements are accomplished in the required frequency band with 28 dB of gain in the coverage region, thanks to the patch optimizations to compensate for phase-delay at several frequencies. To demonstrate this statement, the radiation patterns have also been computed at extreme frequencies assuming ideal reflectarray elements, which provide the same phasing at all the frequencies (Fig. 7.32 for 10.95 GHz). Although the reduction in gain is not significant, as already stated in Chapter 5, the difference in spatial phase delay produces a distortion in the shape of the beam.

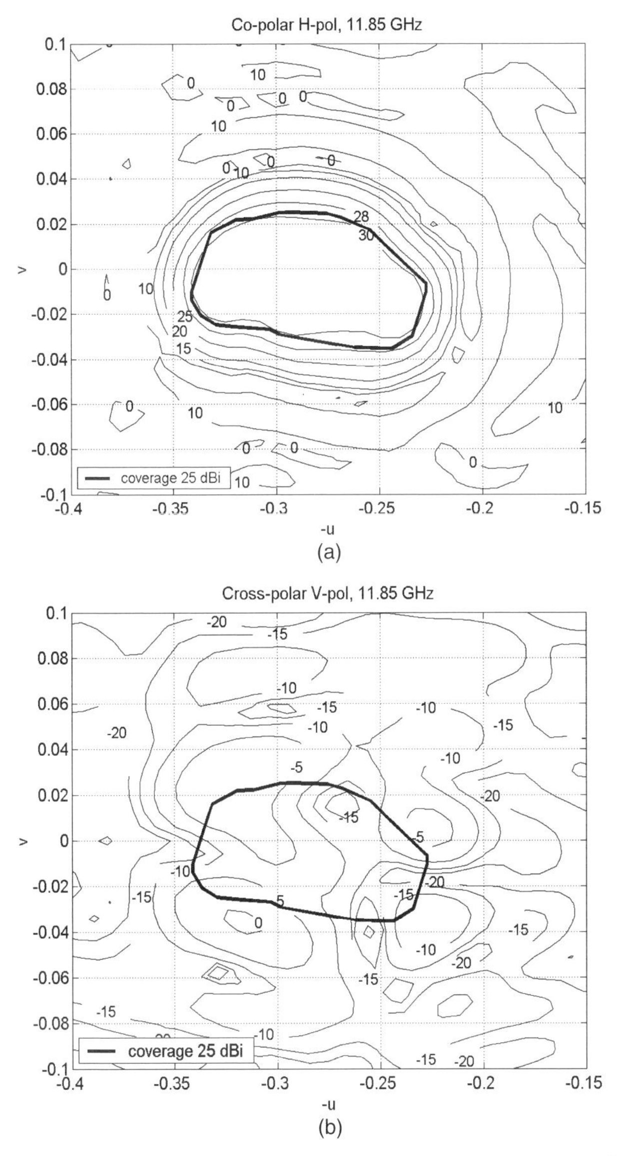

Figure 7.30. Radiation patterns for H-polarization at 11.85 GHz. (a) Co-polar, (b) cross-polar.

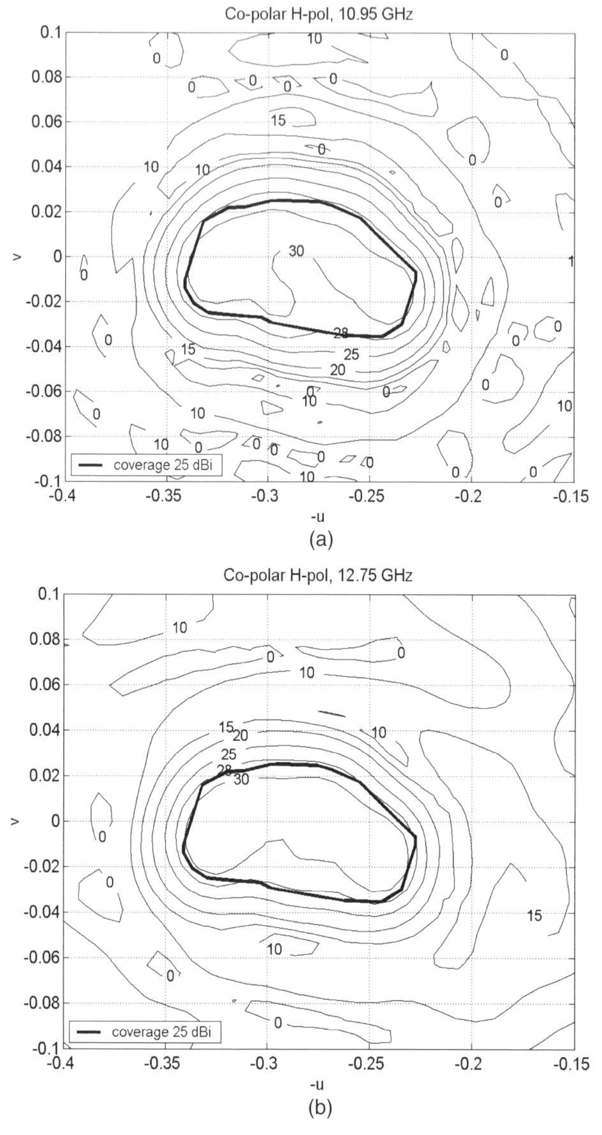

Figure 7.31. Co-polar radiation patterns for H-polarization. (a) At 10.95 GHz, (b) at 12.75 GHz.

The cross-polarization has also been computed, and the results show that 30 dB cross-polar isolation is achieved in 99 percent of the coverage area for a more reduced frequency band (10.95–12.0 GHz). The cross-polar patterns in dBi are shown in Fig. 7.30(b) at 11.85 GHz. The radiation patterns have also been computed considering only the elements inside an ellipse of axes

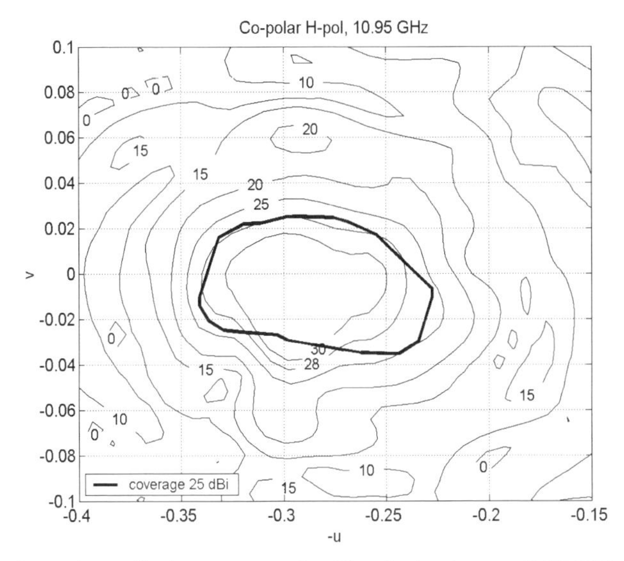

Figure 7.32. Co-polar radiation patterns for H-polarization at 10.95 GHz assuming an ideal phase-shifter element.

1036 × 980 mm (excluding the elements in the corners), and co-polar and cross-polar patterns are practically identical, which confirms that those elements in the corners do not contribute to the radiation pattern, because of the low illumination level.

The results presented in this section have shown the potential use of the reflectarray technology for contoured-beam antennas in space applications. It has been demonstrated that practically any contour requirements can be easily achieved by using the *Intersection Approach* technique, and then by optimizing the patch dimensions on each array layer to fulfill the requirements in a prescribed frequency band. Provided that the required bandwidth for communications and broadcasting space applications can be achieved, multilayer printed reflectarrays can be a technological alternative to onboard shaped reflectors, which reduces costs and manufacturing time, as a result of the elimination of the custom molds used in reflector technology.

7.3 MULTI-BEAM REFLECTARRAYS

Reflectarrays can be designed to generate multiple beams, either with a single feed, or with one or more feeds per beam. In the latter case, the antenna configuration and the design process is similar as in multi-feed reflectors. However, the flexibility provided by reflectarrays to achieve any value of phase-shift independently for each linear polarization, can be used to improve the performances in multi-beam applications, to produce various simultaneous shaped beams, and to achieve several beams with a single feed, as it is described in this section.

First, the problem of generating several simultaneous beams with a single feed is addressed. The required field distribution on a reflectarray to generate several beams can be obtained by simple superposition of the aperture fields associated with the individual beams. Then, to radiate several pencil beams in the directions (θ_n, φ_n), n being an integer number from 1 to the number of beams (N), the field on the reflectarray surface can expressed as

$$E_R(x_i, y_i) = A_R(x_i, y_i)e^{j\Phi_R(x_i, y_i)} = \sum_{n=1}^{N} A_n(x_i, y_i)e^{j\Phi_n(x_i, y_i)}, \tag{7.2}$$

where (x_i, y_i) are the coordinates of the center of element i in the reflectarray, $A_n(x_i, y_i)$ and $\Phi_n(x_i, y_i)$ are, respectively, the amplitude and phase on element i to radiate a beam in the direction (θ_n, φ_n). The amplitude $A_n(x_i, y_i)$ can be considered uniform or tapered, while the phase distribution $\Phi_n(x_i, y_i)$ is a progressive phase, as is known from phased array theory,

$$\phi_n(x_i, y_i) = -K_0 \sin\theta_n \cos\varphi_n x_i - K_0 \sin\theta_n \sin\varphi_n y_i. \tag{7.3}$$

After performing the summation of the complex field distributions, the overall required amplitude and phase distributions $A_R(x_i, y_i)$ and $\Phi_R(x_i, y_i)$ in (7.2) are obtained. The phase distribution corresponding to the multiple beams can be implemented by simply adjusting the phase-shift at each element as in the case of a single-beam reflectarray. On the contrary, the required amplitude distribution exhibit variations on the reflectarray surface, which cannot be achieved with a conventional feed-horn. When the reflectarray is designed to produce the multi-beam phase distribution, but with a smooth amplitude variation imposed by the radiation pattern of the feed, normally the beams are achieved in the required directions, but with high side lobes produced by the difference in amplitude. As a particular case, the beam directions (θ_n, φ_n) can be chosen very close to each other to produce beam overlapping, then a contoured beam is obtained with a single feed, but again the level of the side lobes is normally high. In this case, the multi-beam phase distribution $\Phi_R(x_i, y_i)$ can be used as the starting point in a process of phase-only pattern synthesis for contoured beam reflectarrays, in which the beam shaping will be improved and the side lobes reduced, as already described in Section 7.2.1. In some particular cases, as in the following example, a good multi-beam performance can be achieved by an appropriated design.

In this example, two symmetrical beams on the YZ plane at +27.5° and −27.5° from the Z axis at 11.95 GHz are obtained with a single feed placed on the XZ plane, as shown in Fig. 7.33(a). The required field on the reflectarray surface, given by expression (7.2), in this case becomes

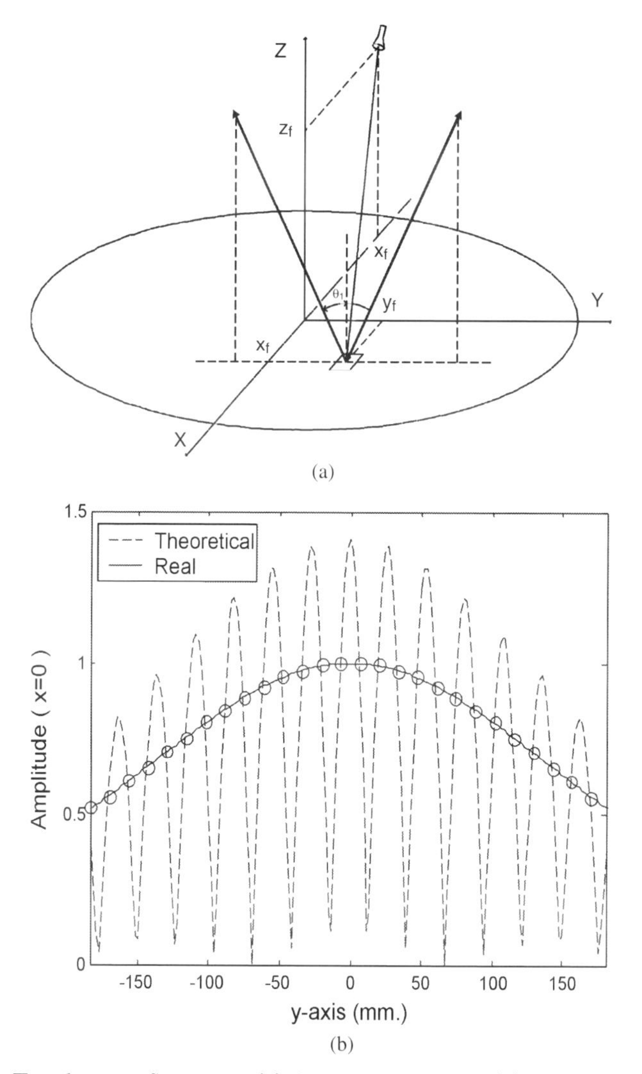

Figure 7.33. Two-beam reflectarray. (a) Antenna geometry, (b) amplitude distribution in the y-direction.

$$E_R(x_i, y_i) = 2A(x_i, y_i)\cos(K_0 \sin(27.5°)y_i), \tag{7.4}$$

which is real as a result of the symmetrical beams. $A(x_i, y_i)$ is the amplitude associated with the individual beams, which are assumed identical. The absolute value of this expression presents the periodical variations in the y-direction associated with the cosine function, as shown in the dashed line of Fig. 7.33(b). The required amplitudes for the points separated by a semi-period

(13.6 mm) are marked with circles. For the reflectarray design, the periodic cell has been chosen as 13.6 × 13.6 (half a period of the oscillations in amplitude), and the radiation pattern and position of the horn have been defined so that the amplitude of the incident field on the reflectarray, shown as a continuous line in Fig. 7.33(b), matches the values marked with circles.

Once the phase distribution and the feed position have been defined, a two-layer reflectarray has been designed, manufactured, and measured. The sandwich consists of two layers of metallic rectangular patches printed on Cuclad ($\varepsilon_r = 3.4$) substrate 250-microns thick and a ground plane, separated from each other by 3-mm Rohacell ($\varepsilon_r = 1.067$) layers. The breadboard is shown in Fig. 7.34. The measured radiation patterns for E polarization (tangential electric field on the reflectarray in the x-direction) are shown in Fig. 7.35, which agrees very well with the simulated patterns, as can be seen in Fig. 7.35(b) for the plane $x = 0$. The patterns have been also simulated and measured at 11.7 and 12.2 GHz, showing a deviation in the beam directions of ±0.7°, and a reduction in gain of less than 0.4 dB, in the whole frequency band. The reflectarray was designed for dual linear polarization and similar patterns were obtained for H-polarization. The dissipative losses measured in the reflectarray were 0.35 dB.

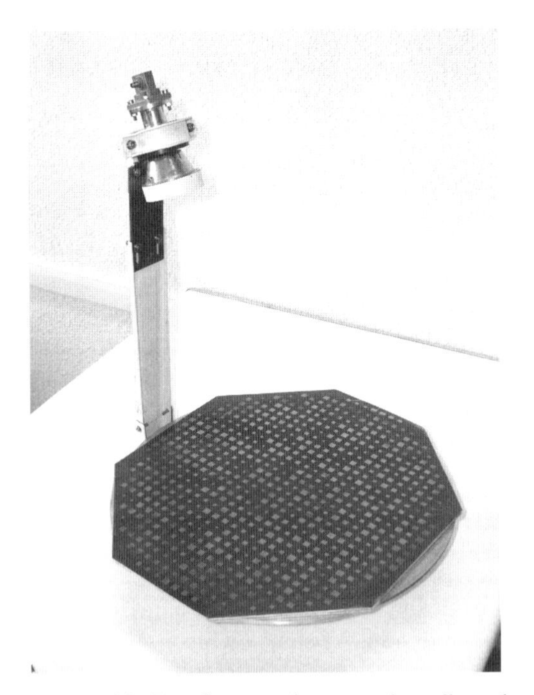

Figure 7.34. Two-beam reflectarray breadboard.

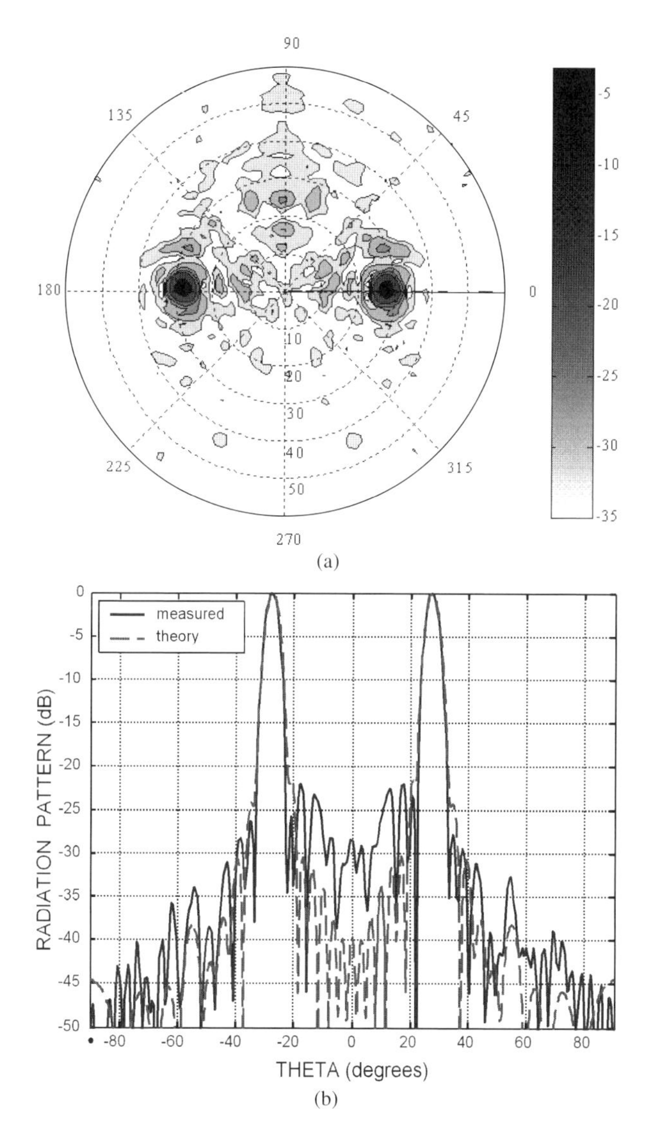

Figure 7.35. Measured radiation patterns at 11.95 GHz for x-polarization. (a) 3-D patterns, (b) patterns on the plane $x = 0$.

For multi-beam applications, in which several simultaneous beams must be generated independently, such as those used in communications with frequency reuse, a reflectarray can be designed to generate a beam in a given direction for each feed, in a similar way as in multi-feed reflectors. The reflectarray is designed to simulate the electrical behavior of a parabolic reflector that generates a collimated beam in the direction (θ_0, φ_0) for a feed located at the focal point. When other feeds are placed in the vicinity of the focus, several beams are generated in the directions (θ_n, φ_n), with subindex n varying from 1 to the number of feeds, but some aberration effects occur in the radiation patterns for the surrounding beams due to the feed defocusing. This problem has been studied in parabolic reflectors [25], and optimal feed positions can be found to minimize the aberration. For further improvements, not only the feed positions but also the patch dimensions on the reflectarray can be optimized.

The next example is a multi-fed reflectarray with four beams for X-band Synthetic Aperture Radar (SAR) applications in microsatellites. The proposed multi-beam configuration consists of a 1.6-m offset reflectarray with four feeds and an f/D around one, as shown in Fig. 7.36, which was designed for dual linear polarization. The four beams, called SS1, SS2, SS3, and SS4, are radiated in the XZ plane in the directions with respect to the Z_r axis of $\theta_b = 20.91°$, 18.41°, 15.91°, and 13.41°, respectively, and each one is related to only one feed placed on the same plane. The reflectarray is designed for the beam SS3 with the corresponding feed at coordinates $x_{f3} = -540$, $y_{f3} = 0$, and $z_{f3} = 1714$ mm, for the central frequency of 9.65 GHz, then the rest of the feeds are positioned to radiate the beam in the required directions. The radiation patterns in gain for the four beams at 9.65 GHz are superimposed in Fig. 7.37 in azimuth and elevation planes for y-polarization, being very similar for the orthogonal polarization. Gain and beamwidth are practically the same for the four beams because of the relatively large f/D ratio. It has been checked that the radiation patterns remain without distortion in the required frequency band for the X-band SAR, which is only 2.3 percent. The simultaneous beams can be shaped, by applying the *intersection approach* method, then adjusting the phases for one of the

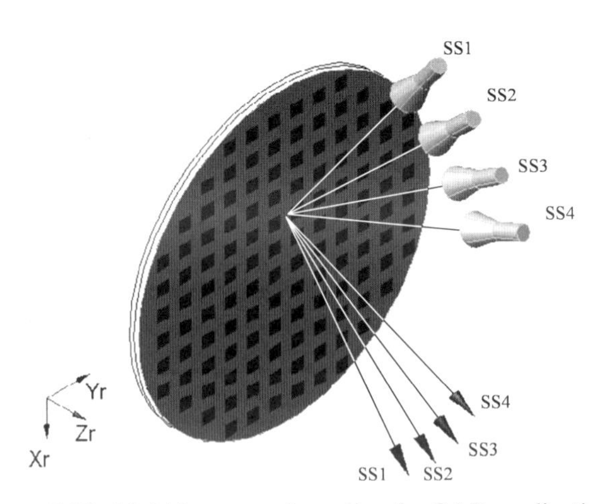

Figure 7.36. Multi-beam configuration for SAR applications.

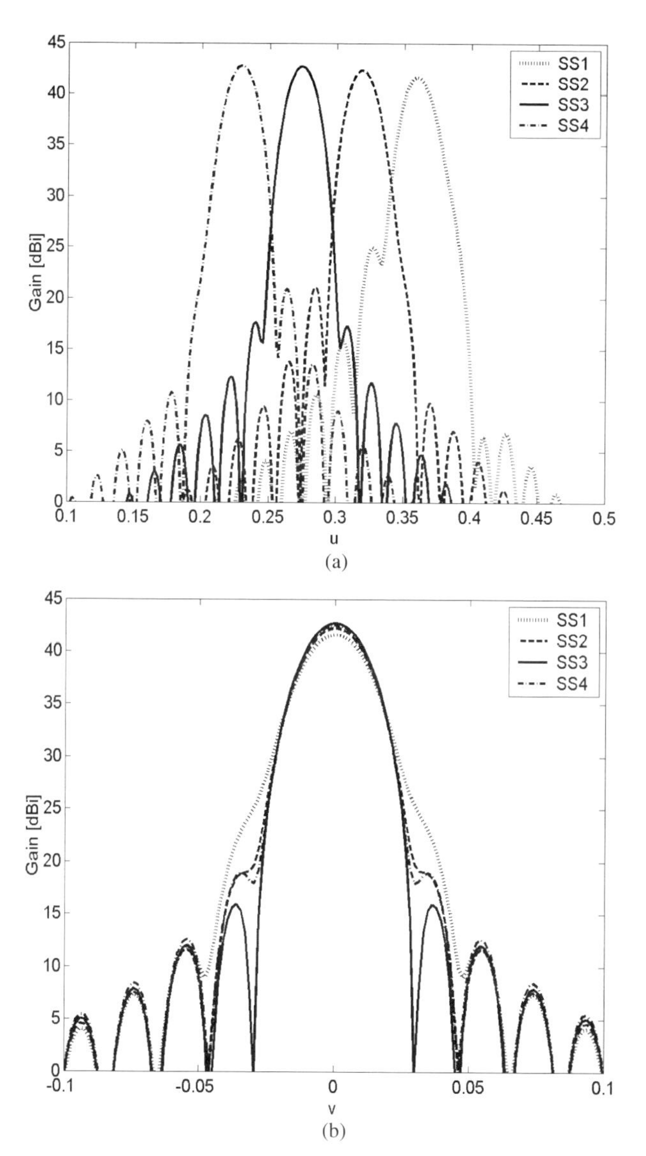

Figure 7.37. Comparison of the four beams for y-polarization. (a) Elevation plane, (b) azimuth plane.

central beams, and finally positioning the rest of the feeds for the other beams, as shown in the next application.

The last example shows a multi-fed and multi-beam reflectarray designed for a central station sectored antenna for point-to-multipoint communications (LMDS) at 25.5 GHz [26, 27]. The antenna is designed to generate three independent shaped beams, covering adjacent 30° sectors in azimuth with the same cosecant-square pattern in elevation. Each beam is generated by a feed that illuminates a square two-layer reflectarray made of 30×30 elements (Fig. 7.38(a)). The periodic cell of dimensions $5.84\,\text{mm} \times 5.84\,\text{mm}$ is the same as shown in Fig. 7.13 but with only two stacked patches. The patches are printed on 0.79-mm-thick CuClad 233LX and the relative dimensions between stacked patches is $a_1/a_2 = b_1/b_2 = 0.75$.

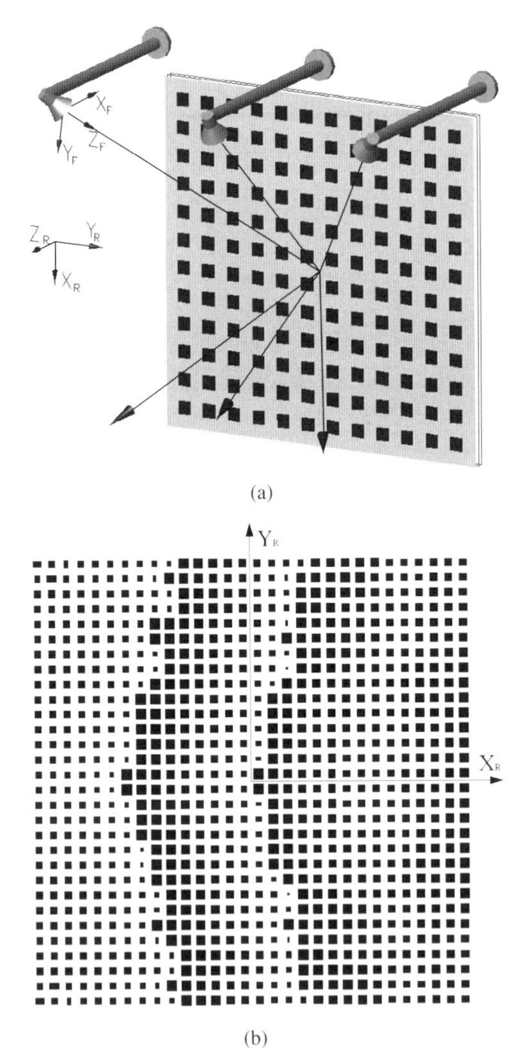

Figure 7.38. Reflectarray for three shaped beams. (a) Antenna configuration, (b) layout for the second layer.

A commercial pyramidal horn (FMI 20240 from Flann Microwave Ltd.) has been selected for the feeds. In the far-field region, the horn can be modeled as a $cos^q(\theta)$ function with a q-factor of 37, but the reflectarray elements are placed in the Fresnel region of the feed-horns. For a more accurate design, the real near-field incident on each reflectarray cell obtained from measurements of the horn as described in [26, 27] is considered. The central feed is located at coordinates $x_{cf}=-94$, $y_{cf}=0$, $z_{cf}=214$ (in mm), referred to the center of the reflectarray. For the reflectarray design, first the pattern synthesis technique described in Section 7.2.1 is applied to obtain the required phase distribution to generate the shaped central beam associated to the central feed at 25.5 GHz. In this process, the phase distribution is enforced to be symmetric with respect to the X_R-axis, in order to have symmetrical performance for both lateral beams. Then, the patch dimensions are adjusted to produce the required phase distribution for dual linear polarization as described in [28]. Fig. 7.38(b) shows the layout of the second array layer (closer to the ground plane), while in the first layer the patches are scaled by 0.75. Finally, the symmetrical positions of the lateral feeds are optimized to generate the shaped beams with the appropriate direction and minimum distortion, and the resulting coordinates are $x_{lf}=-94$, $y_{lf}=\pm113$, $z_{lf}=182$ (in mm).

Although the reflectarray has been designed for dual polarization, alternate polarizations are considered in the following results, using vertical (V) and horizontal (H) polarization for the lateral and central beams, respectively. The radiation patterns are computed for the central and lateral beams, considering the real incident field on the reflectarray elements and using the analysis routine based on MoM. The 3-D patterns are shown in Fig. 7.39 for the central beam in H-polarization (a) and the lateral one in V-polarization (b). Note that only one lateral beam is plotted, the other one being symmetric because of the antenna symmetry. The radiation pattern for the central beam is compliant with the requirements, as shown in Fig. 7.40 for the principal planes.

Although a good shaping is obtained for the lateral beams (Fig. 7.39(b)), the mask of requirements are not completely satisfied, as shown in Fig. 7.40. If alternate polarization is used for central and lateral beams, a further improvement in the shaped lateral beams can be achieved by optimizing the lengths of the patches in only x-dimension, for example, for the V-polarization associated to the lateral beams. These results show that several shaped beams can be generated in reflectarray technology in a one-feed-per-beam basis.

7.4 AMPLIFYING REFLECTARRAY

It has been recognized that, in addition to the replacement for a passive parabolic reflector, a reflectarray can also act as an amplifying array, as well as a spatial power combiner, by including amplifiers in individual phased microstrip patch antenna elements. Bialkowski [29] of the University of Queensland, Australia, has demonstrated both of these reflectarrays techniques. Both

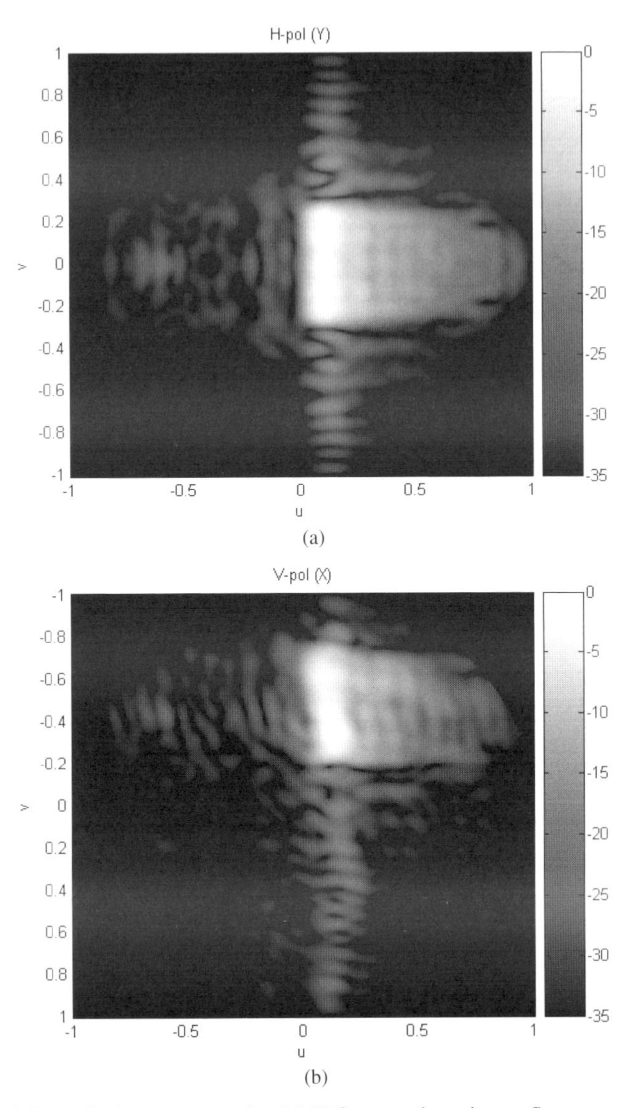

Figure 7.39. 3-D radiation patterns for LMDS central station reflectarray. (a) Central beam, (b) lateral beam.

reflectarrays were constructed and tested in the frequency range of 9 to 11 GHz. The amplifying reflectarray having 137 dual-polarized patch elements is shown in Fig. 7.41. A feed horn illuminates these elements with one polarization, and upon reception, each patch sends a signal via a coupling slot through an amplifier and reradiates out into space from its orthogonal port, as illustrated in Fig. 2.9 in Chapter 2. The required phase of each element for the reflectarray

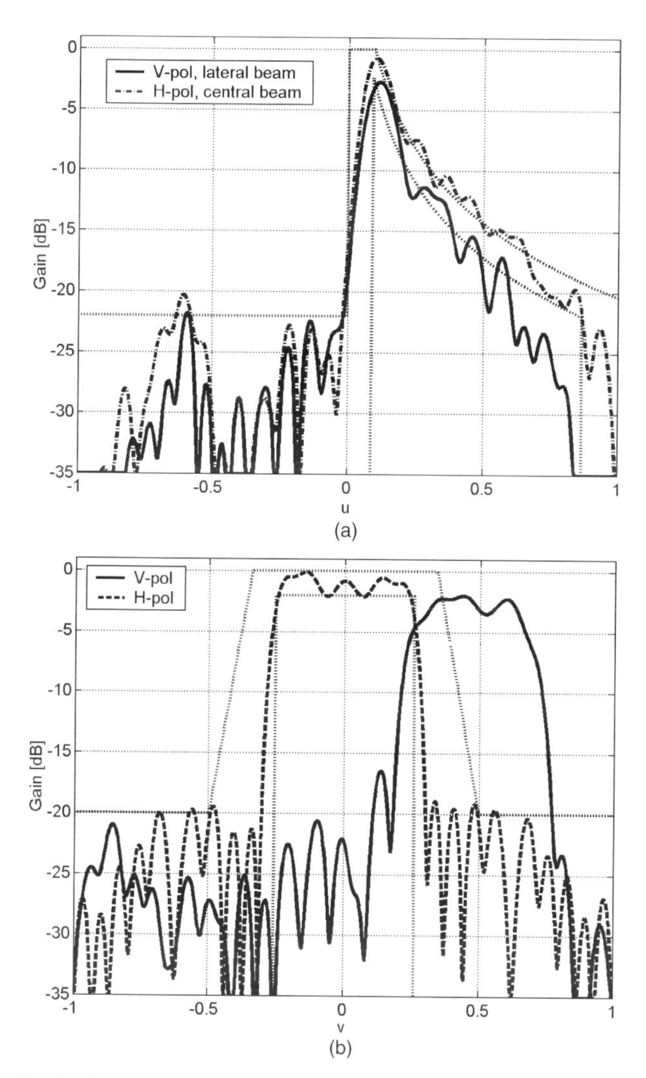

Figure 7.40. Radiation patterns. (a) Elevation pattern [dashed line at plane v=0° and continuous line at plane v=sin(30°)], (b) Azimuth pattern [plane u=sin(5°)].

to radiate a co-phasal plane wave is achieved by varying the length of the transmission line between the two orthogonal ports. Each amplifier, using MESFET technology, has a gain of 10 dB across the frequency range of interest. The reflectarray did achieve a focused main beam in the designed broadside direction with a power gain of about 10 dB.

The second antenna is a spatial power combining reflectarray with 37 elements, as shown in Fig. 7.42. Its feed array, shown in Fig. 7.43, has four microstrip

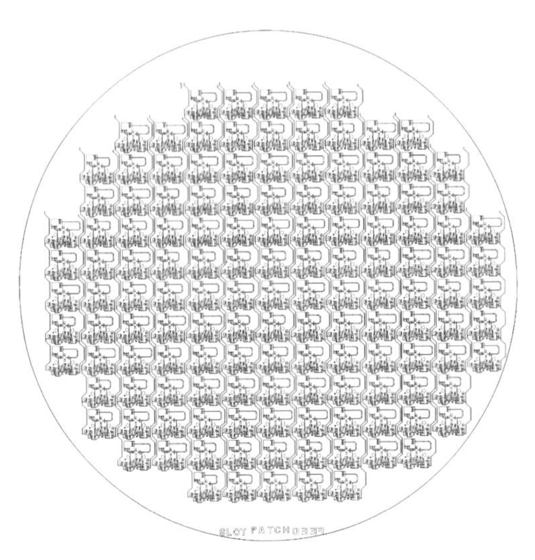

Figure 7.41. 137-element microstrip amplifying reflectarray. (Courtesy of Professor Bialkowski, University of Queensland, Australia.)

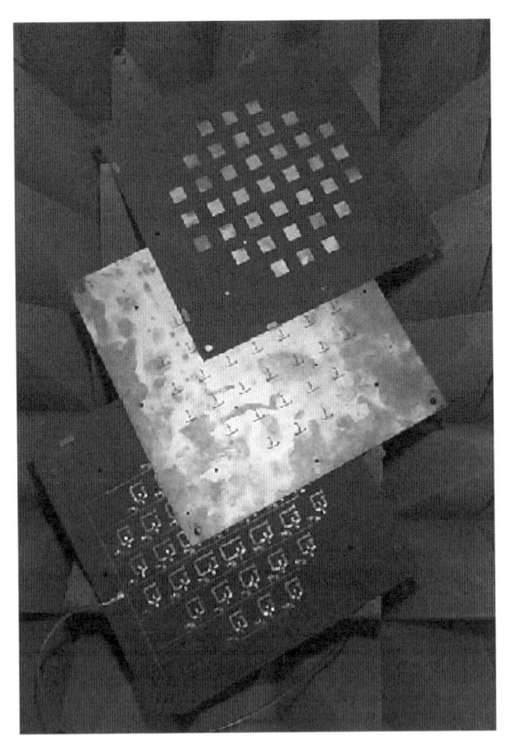

Figure 7.42. Photograph showing three layers (patch, slot, and amplifier) (Courtesy of Professor Bialkowski, University of Queensland, Australia.)

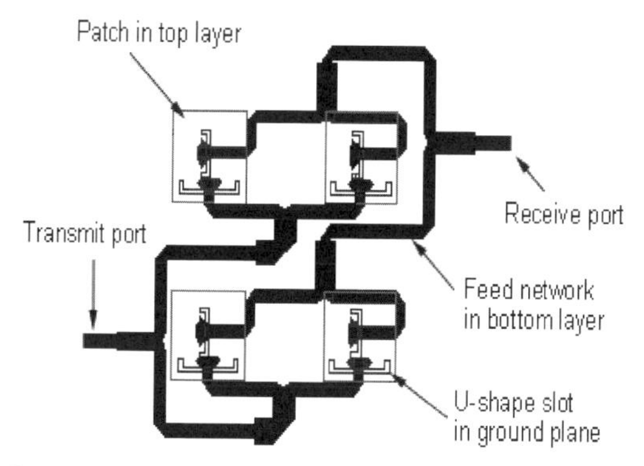

Figure 7.43. Feed array with four patches that transmit vertical polarization while receive horizontal polarization. (Courtesy of Professor Bialkowski, Univ. of Queensland, Australia.)

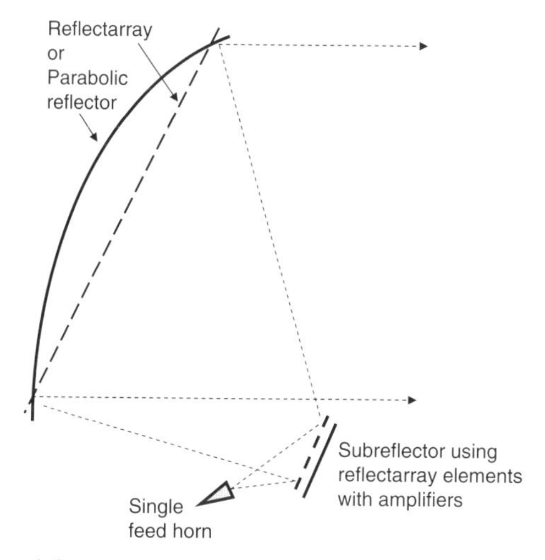

Figure 7.44. Proposed Cassegrain configuration with the subreflector being a reflectarray having amplifiers to provide high power.

patches with both transmit and receive capabilities. The vertically polarized transmit feed array sends signals to the 37 reflectarray elements, where they get amplified, their polarization is changed to horizontal, and then reradiated to focus the beam to the horizontally polarized receive array. Thus, through the above process, the signal gets amplified and spatially combined without encounter any significant loss. This power combiner achieved a power gain of about 7 dB.

With the success of the above amplifying reflectarrays, the author (J. Huang) of this book section proposes a Cassegrain configuration, as shown in Fig. 7.44, where the subreflector is of a reflectarray design that can illuminate a very large aperture with either a reflectarray design or a parabolic reflector design. The subreflector elements have amplifiers that allow relatively low-power, solid-state amplifiers to be combined spatially to provide very high overall antenna radiated power with graceful degradation in performance. In doing so, a relatively small number of amplifiers, such as tens or hundreds, can be used on the subreflector, instead of millions on the very large main reflectarray, to provide high power and to minimize complexity and cost.

7.5 FOLDED COMPACT REFLECTARRAY

In reflectarrays, as in reflector antennas, the feed is placed at a certain distance from the reflecting surface, removing the low-profile characteristic of printed antennas. In order to overcome this problem, a more compact configuration called a "folded reflectarray antenna" has been proposed for linear polarization applications [30–32]. The folded configuration consists of a printed reflectarray with a feed embedded in its structure and a polarizing grid placed parallel to the reflectarray at a certain distance h, as shown in Fig. 7.45. The polarizing grid is made of a strip grating printed on a dielectric slab, so that it behaves as a reflector for the electric fields parallel to the grating while it is transparent for the orthogonal polarization. Alternatively, the grating can be substituted by a slot-array. The folded configuration allows a reduction of the antenna depth to approximately one half of the focal distance (Fig. 7.45).

The detailed operation of this antenna has been explained in [32] and is briefly described here. The feed produces an approximately linearly polarized spherical wave with the electric field parallel to the grating, which is reflected by the polarizing grid towards the printed reflectarray. The reflectarray is formed by variable size patches rotated by 45° with respect to the direction of incident electric field (Fig. 7.45(b)). The reflectarray is designed to properly focus or shape the beam and to twist the polarization by 90° with respect to the incident wave, so that the radiated field can pass through the polarizer. The polarization rotation is produced by adjusting the patch dimensions to cause a phase difference of 180° between the reflection phases of the two incident wave components, as schematically shown in Fig. 7.46. The polarizing radome is designed with the grating printed on a dielectric substrate, whose thickness is approximately half a wavelength, to allow the field radiated by the reflectarray to pass through with minimum insertion losses.

If a slot-array is used as the polarizer, an appropriate design taking into account the substrate thickness must be carried out to minimize the reflection losses.

The phase-shift on the reflectarray is controlled by adjusting the two dimensions of rectangular patches printed on a grounded substrate. The use of two dimensions provides two degrees of freedom in the design for linear polarization. One allows for adjusting the phases for the focusing or beam-shaping requirements, so that a spherical wavefront produced by the feed is

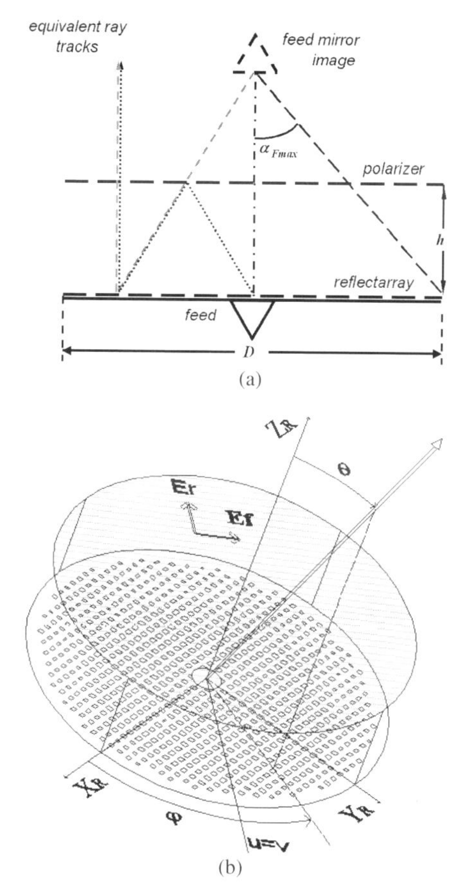

Figure 7.45. Folded printed reflectarray. (a) Lateral view with the equivalent feed, (b) 3-D drawing.

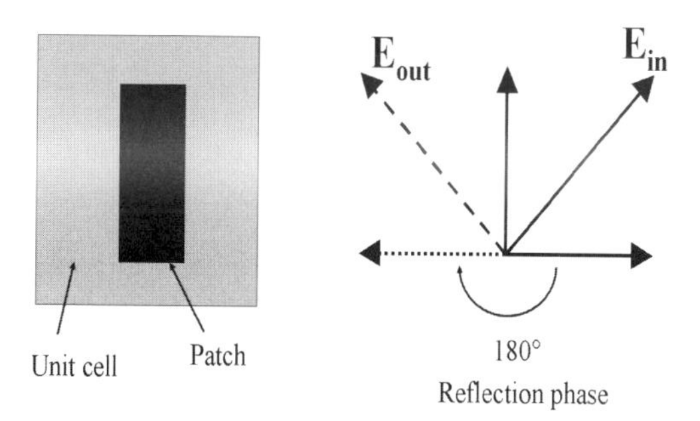

Figure 7.46. Illustration of the orthogonal components of the incident and reflected electric field for a difference of 180° in the reflection phases.

transformed into a planar wavefront, or into a prescribed shaped beam. The other one allows for obtaining a 180° phase difference between the two orthogonal polarizations, in order to produce the polarization twist.

For the analysis and design of the folded antenna, the polarizing grid is assumed ideal and the equivalent reflectarray with the feed placed at the specular position is considered (Fig. 7.45(a)). Note that the folded reflectarray provides very low cross-polarization levels even for lower f/D ratios; first, because it is a centered-fed configuration, and second, because the cross-polarization is filtered by the polarizer grid. For a low-profile compact antenna, the height h should be reduced as much as possible, and so the focal distance. However, the maximum angle of incidence α_{Fmax} should not exceed 40–45°, in order to avoid problems in the electrical design (when the angle of incidence is large, the phase-shift is different for the two polarizations and also differs from the normal incidence values, as shown in Chapter 3). Hence, for practical cases, the antenna height can be reduced to a maximum of one quarter of the diameter.

Taking into account that reflectarrays are usually designed for a diameter of at least ten wavelengthrs, the folded configuration has a big potential in Ka-band and higher frequencies, where the antenna depth can be reduced to a few centimeters. A further reduction in thickness can be achieved by a triple-folded configuration, as described in [33] at the cost of a small degradation on the electrical performances. Millimeter-wave folded printed reflectarrays have been proposed for radar and communication applications in [31, 32], where compact and robust antennas are required. High-gain, omni-directional, sectorial-beam, mechanically steerable, and multi-feed, multi-beam antenna solutions have been successfully demonstrated for different configurations of folded reflectarray antennas.

A multi-fed folded reflectarray antenna with multiple beams was presented for the first time in [34]. A prototype with the polarizer radome taken apart is shown Fig. 7.47. The same antenna can be used to scan the beam by switching from one feed to the next. The multi-beam antenna was designed on the basis of one feed per beam, using the principle of bifocal antennas and a ray tracing technique.

In principle, the folded antenna is designed for a low focal distance to keep a low profile, and consequently the scanning capabilities will be poor. However, the concept of bifocal antennas has been used to design lens or reflector antennas with large scanning angles [35, 36]. In bifocal antennas, a focal ring replaces the focal point by using an additional degree of freedom, which is implemented by shaping both surfaces on the lens antenna [35], or both reflectors on a double reflector configuration [36]. The additional degree of freedom in the folded configuration is achieved by adjusting the phase of the reflected field at the polarizer radome. For that purpose, the polarizing grid acts as the ground plane of a new reflectarray for the polarization of the incident field, which is based on varying-sized dipoles parallel to the grid so that the reflectarray is still transparent for the orthogonal polarization (Fig. 7.48).

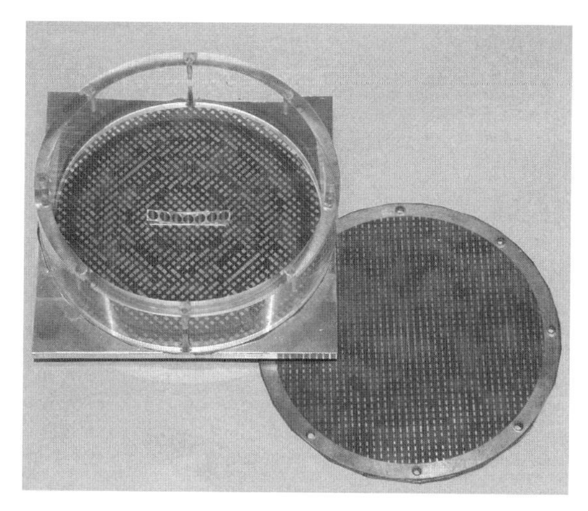

Figure 7.47. Prototype of folded reflectarray for multi-beam operation. (Courtesy of Professor Menzel of Univ. of Ulm, Germany.)

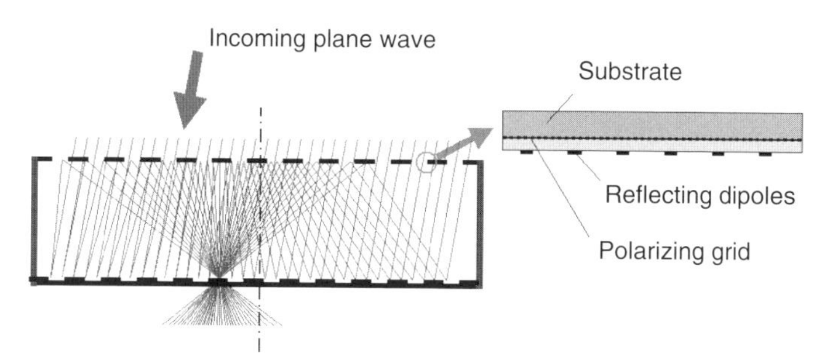

Figure 7.48. Cross-section of bifocal folded reflector antenna and lay-up of polarizer radome.

Considering the independent phase adjustment on both reflectarray surfaces, the folded reflectarray can be designed using the principle of the bifocal antenna, where the required phase-shift on each surface is determined by a ray tracing technique, in a similar manner as in [35, 36]. A folded reflectarray with nine beams at 37.5 GHz was designed, manufactured, and tested recently [37]. The two reflectarrays of diameter 180 mm are printed on a 0.5-mm-thick substrate ($\varepsilon_r = 2.2$) and separated 50 mm. Open rectangular waveguides separated 5.1 mm in the E-plane (center to center) are used as feed elements. The measured radiation patterns in the principal planes are shown in Fig. 7.49.

Folded reflectarrays have also been successfully demonstrated for shaped beam applications. As a first example, a 40° sector beam antenna with narrow beam in the orthogonal plane for 37.5 GHz has been reported in [37]. As in

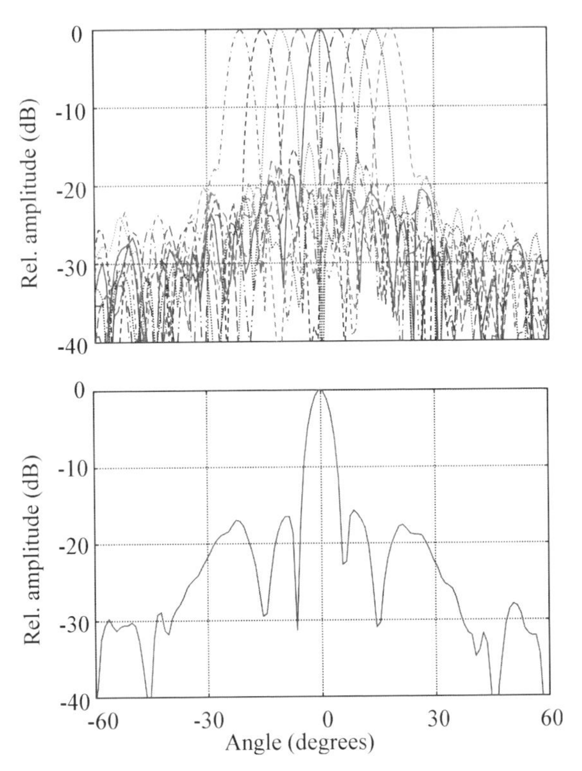

Figure 7.49. Experimental H-plane (top) and E-plane (bottom, central beam only) radiation diagrams for a nine-beam reflectarray antenna (frequency 37.5 GHz). (Courtesy of Professor Menzel of Univ. of Ulm, Germany.)

conventional reflectarray antennas, the amplitude distribution is fixed by the feed radiation pattern, and the beam shaping is achieved by implementing on the reflectarray a phase-shift distribution obtained by a phase-only synthesis technique.

The phase distribution on the reflectarray surface is obtained by combining the linear phase distributions in the principal planes. The phase corresponding to a sector beam in the azimuth plane was obtained by the technique described in [38], while the phase corresponding to a pencil beam is chosen in the elevation plane. The antenna diameter and distance between reflector and polarizer are, respectively, 150 mm and 40 mm. The measured radiation patterns show a good performance in a 2 GHz band (5 percent), with beamwidths in azimuth and elevation planes nearly 40° and 3.5°, respectively. By combining the two previous approaches of multi-beam bifocal antenna and phase-only synthesis, a folded reflectarray with three simultaneous 30°-sector beams in azimuth has been designed, manufactured, and measured, as reported in [39].

Another approach to generate shaped beams in a folded configuration with a second reflectarray in the polarizing grid consists of using the phase adjustment in the polarizer layer to redistribute the energy from the feed. For example, an ideal sector-beam is achieved when the amplitude distribution on the main reflectarray is a *sinc* function. Then, the top reflectarray on the polarizer layer (Fig. 7.48) is designed by ray tracing to produce a sinc amplitude distribution on the azimuth plane. The resulting radiation patterns given in [40] shows a sector beam with an abrupt roll-off outside the sector.

Previous folded reflectarrays were made of a single layer of varying-sized patches, and therefore they suffer from the bandwidth limitation inherent to reflectarrays. In order to improve the bandwidth, a three-layer folded reflectarray has been designed in the 24.5–26.5 GHz band for a LMDS central station antenna, as reported in [41]. This application requires a shaped pattern of cosecant squared beam in elevation and a sector beam in azimuth. The design method for shaped-beam reflectarrays, described in [10], has been conveniently modified to be used in the folded configuration. First, the phase-only synthesis technique known as *Intersection Approach* [14] is applied to achieve the required phase-shift on the reflectarray elements that gives the desired shaped radiation pattern at the central and extreme frequencies, as described in [15] and [23]. Then, the stacked printed elements of the reflectarray are designed individually, in order to achieve the previously synthesized phase-shift distribution at the central and extreme frequencies, and to ensure that the reflected field on each element is orthogonally twisted. This second fact is implemented by enforcing a phase difference of 180° between the two orthogonal components of the reflected field (associated with each patch dimension) when optimizing the dimensions of the stacked rectangular patches. More details of the design process are given in [41].

The final design practically meets the requirements for a LMDS base station antenna, defined as a 6°-tilt cosecant squared beam in elevation and a 90°-sector beam in azimuth. The reflectarray is made up of three-stacked arrays of rectangular patches printed on RT Duroid 5870 0.508-mm thick arranged in a 5.85 mm side square lattice. The diameter of the reflectarray and polarizing grid is 18 cm and they are separated by a distance of 5 cm. The manufactured breadboard is shown in Fig. 7.50 with the polarizer taken apart to show the upper array of patches, and the 3-D co-polar measured patterns at 25 GHz are shown in Fig. 7.51. The simulated and measured gain patterns in elevation and azimuth planes at 25 GHz are shown in Fig. 7.52 superimposed to the requirements' mask. The slight differences between simulated and measured gain patterns are produced by the effect of the finite dimensions of the polarizer. Since the reflectarray was designed for a wide beam (90° in azimuth), part of the energy is being radiated near and even beyond the polarizer edges, and the gain suffers from the diffraction effects. It was checked that the beam shaping and side-lobe performances were improved by increasing the polar-

Figure 7.50. Manufactured three-layer folded reflectarray for LMDS base station coverage.

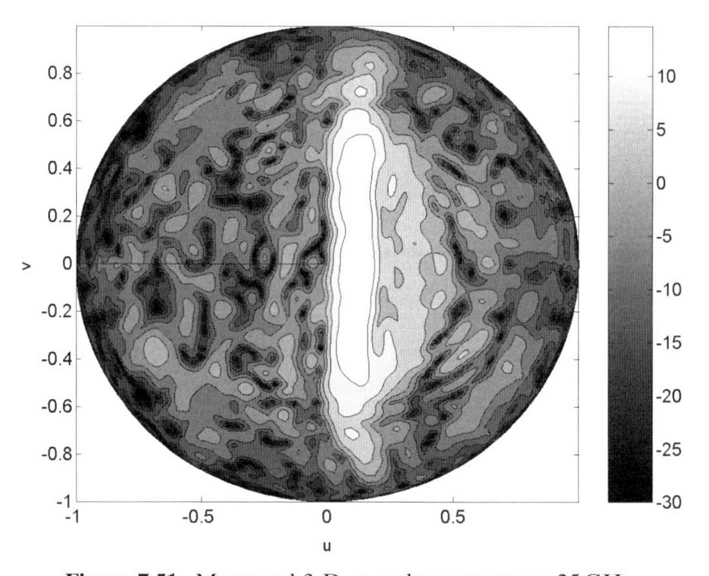

Figure 7.51. Measured 3-D co-polar patterns at 25 GHz.

izer dimensions. Although the antenna was designed for the 24.5–26.5 GHz band, the measured radiation patterns exhibit a shaping close to the requirements in the 24–26 GHz frequency band, showing a 500 MHz shift down in frequency, which is produced by tolerance errors. The errors measured in the patch dimensions are from 20 to 100 microns in excess of the nominal dimensions, justifying a frequency shift towards lower frequencies and some distor-

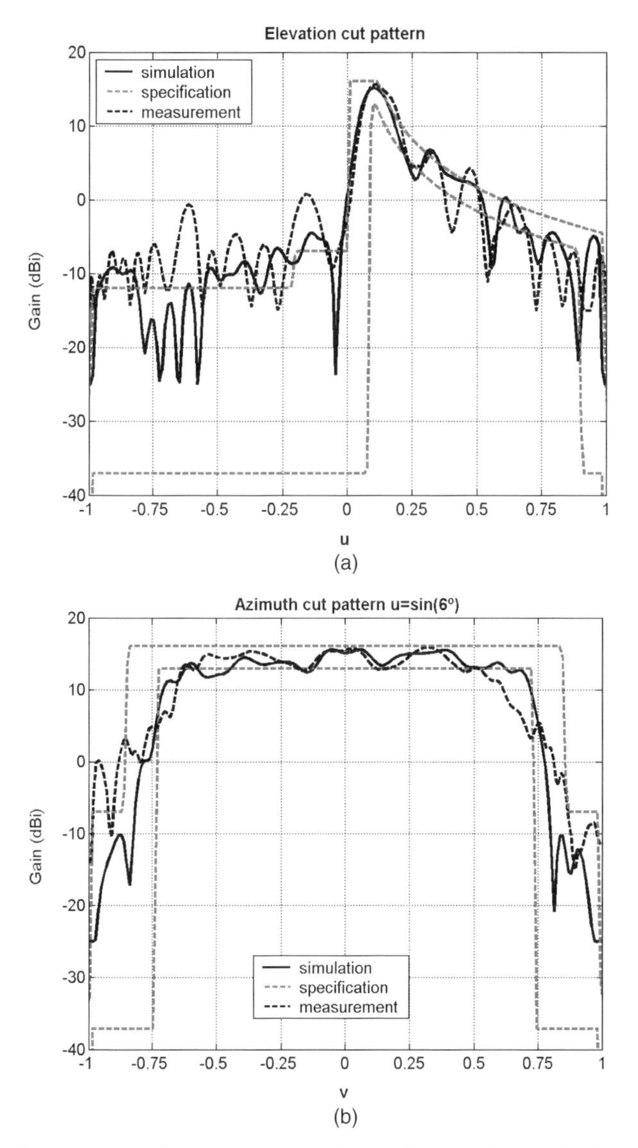

Figure 7.52. Comparison of measurements, simulations, and specifications at 25 GHz. (a) Elevation plane pattern v = 0; (b) azimuth plane pattern u = sin(6°).

tions on the patterns. However, these errors can be significantly reduced since a very good accuracy can be achieved in the photo-etching process (better than ±20 microns). The results shown in this section demonstrate the viability of folded reflectarrays for multi-beam and shaped-beam antennas in a bandwidth up to 8 percent.

7.6 CASSEGRAIN OFFSET-FED CONFIGURATIONS

The reflectarray, which has similar optics behavior as that of the parabolic reflector, can be designed with a dual-reflector configuration using either a Cassegrain or a Gregorian approach [42]. There are several advantages associated with the dual-reflector configuration. First, due to the folded optics, the overall antenna profile height can be reduced. Second, the offset feed can be adapted to eliminate the blockage effect introduced by the feed system. Third, with a small phased array placed close to the subreflector (subreflector in the near-field distance of the phase array), the main beam can be made to achieve small angle scan as the phases of the feed array are changed [43]. This is often desired for very large apertures in order to achieve fine beam pointing. Finally, when compared with a single reflector, the dual-reflector system can achieve a much longer equivalent focal length by specially designing the subreflector. With a longer focal length, the bandwidth limit of the reflectarray can be relaxed [44].

The Cassegrain and Gregorian configurations are basically the same with the difference being that the Cassegrain generally uses convex hyperboloid subreflector, while the Gregorian uses concave ellipsoid subreflector. From here on in this section, only Cassegrain design will be discussed, with the knowledge that Gregorian can also be used.

There are several Cassegrain configurations that can be applied by employing the reflectarray as the primary aperture. The first one, shown in Fig. 7.53, is a center-fed Cassegrain design where the subreflector is a hyperboloid and the blockage loss is generally unavoidable. For reflectarray application, the

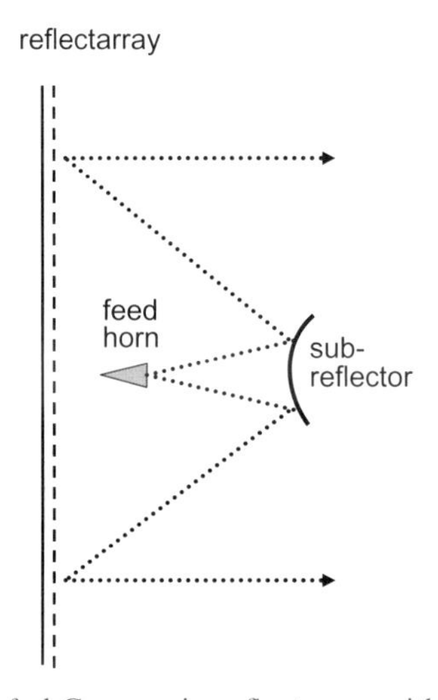

Figure 7.53. Center-fed Cassegrain reflectarray with a single feed horn.

incident angles (angle from the broadside direction) from the feed to the reflectarray elements are generally smaller when compared with the offset-fed design, which can yield better efficiency (discussed in Chapter 4). This configuration is better suited for large aperture applications where the subreflector has a smaller area ratio to the main aperture.

The second configuration, shown in Fig. 7.54, is an offset-fed Cassegrain design where the blockage effect is minimized. Notice that it is better to design the out-going beam to have the same subtend angle from the broadside direction as the incident angle. In doing so, as discussed in Chapter 4, the reflected energy and the reradiated energy of each reflectarray element can be co-located in the same direction and not wasted.

The third configuration, shown in Fig. 7.55, is also offset-fed but with a phased array feed. In this configuration with the subreflector placed in the near-field distance of the array, fine beam pointing can be achieved by properly changing the phase front of the feed array. Often in spacecraft applications, where altitude control and on-board gimbals systems cannot point the antenna accurately, the antenna itself must perform the fine beam pointing.

The fourth configuration, shown in Fig. 7.56 and similar to Fig. 7.44, has the offset-placed subreflector also made of reflectarray [45]. With this design, amplifier modules and/or phase shifters can be implanted in all reflectarray elements on the subreflector to achieve high-power transmission and/or fine beam pointing.

The first Cassegrain reflectarray using microstrip patches [46] was developed with a center-fed geometry at 77 GHz. It has a configuration similar to that shown in Fig. 7.53 and has a reflectarray diameter of 15.2 cm (6 inches) with 5770 linearly polarized rectangular patches using the variable-size patch design. The subreflector with a hyperbolic design has a diameter of 1.61 cm. Although the geometry was not optimally designed, the antenna achieved

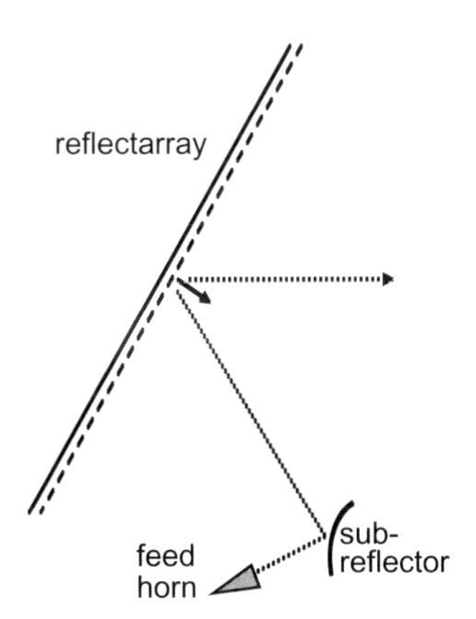

Figure 7.54. Offset-fed Cassegrain reflectarray with a single feed horn, subreflector being a hyperboloid.

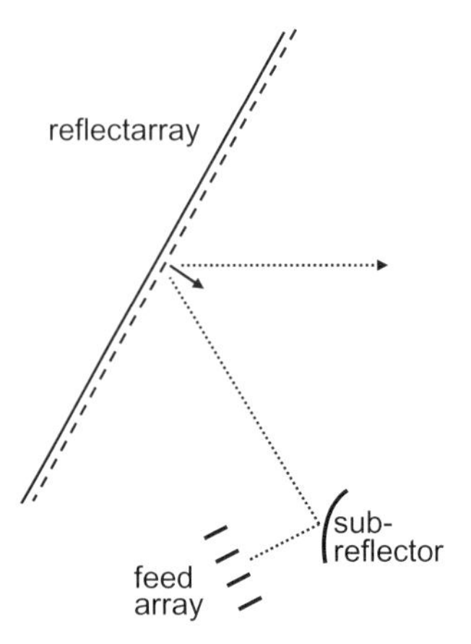

Figure 7.55. Offset-fed Cassegrain reflectarray with a phased array feed for fine beam pointing.

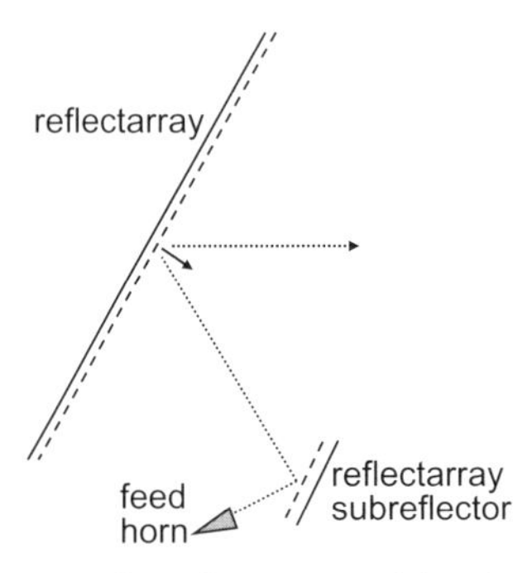

Figure 7.56. Offset-fed Cassegrain reflectarray with subreflector being a reflectarray.

good radiation patterns with a measured gain of 36.9 dB (calculated directivity is 42.8 dB).

The first Cassegrain reflectarray with an offset configuration, similar to that shown in Fig. 7.54, was developed for an X/Ka dual-band performance [47]. This antenna performance was already presented in Chapter 6 and is briefly described here again. It has a main aperture diameter of 75 cm and uses annular ring elements with each element having two capacitive gaps for circular polarization. The main reflectarray aperture consists of two shared-aperture overlapping thin membranes. The X-band membrane having etched larger annular ring elements is situated on top of the Ka-band membrane with smaller annular ring elements. The subreflector, which is a convex hyperboloid, has a projected elliptical aperture with major axis dimension of 150 mm and minor axis dimension of 84 mm. Both the measured X- and Ka-band patterns show excellent performance with both sidelobe and cross-polarization levels below –25 dB. Both bands achieved aperture efficiencies close to 50 percent.

A third development that is worth mentioning here is an analysis tool [48] recently developed to analyze and design the Cassegrain configuration using a phased array feed similar to that shown in Fig. 7.55. It uses a very efficient physical optics (PO) method to solve the subreflector, as well as the main reflectarray fields. The PO current is assigned to the surface of individual reflectarray elements with a correction phase (required phase for beam formation) applied. Although this method is an approximation, with no mutual coupling accounted for between the reflectarray elements, it is very efficient to analyze a very large antenna system with nearly a million elements and tens or hundreds of feed array elements involved. As an example, a 3-m Ka-band (32 GHz) Cassegrain reflectarray having approximately 275,000 elements was analyzed with 5×5 feed array elements. For this study, it was attempted to see how much main beam scan can be achieved with a –3 dB gain loss tolerated if the feed array is phase adjusted. A calculated result, shown in Fig. 7.57, demonstrated that, with the feed array adjusted for a 15° beam scan, the resultant reflectarray beam scan achieved is 0.3°. This beam scan is achieved by placing the subreflector in the near-field region of the phased array. Tilting the planar wave front of the phased array is equivalent to moving the array's phase center [43].

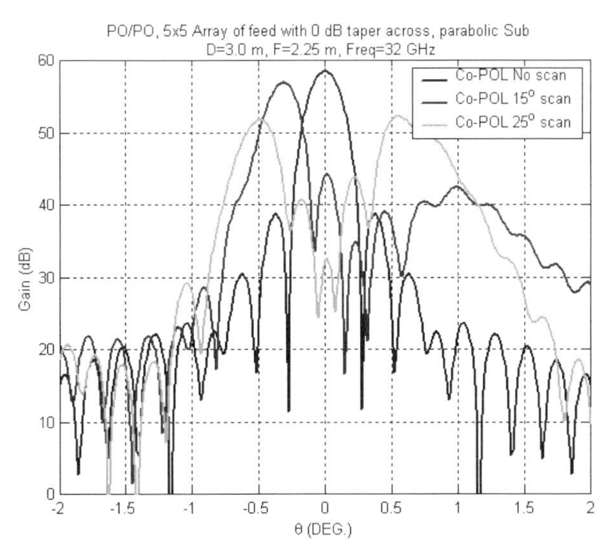

Figure 7.57. Calculated patterns of a 3-m Ka-band Cassegrain reflectarray with feed array scanned to 15° and 25°.

7.7 VERY LARGE APERTURE APPLICATIONS

As discussed in Section 7.1, due to the modern-day need for longer distance communications, higher data rate, and finer radar imaging resolution, one of the few solutions is to have a higher-gain or large-aperture antenna. For NASA's space programs, telecommunication antennas with sizes in the range of 1-4 m have previously been used in the frequency bands up to Ka-band [49], and Earth remote sensing antennas with sizes up to 10 m have been employed in the frequency range from L- to X-band [50].

For future systems, antenna sizes in the range of 10 to 100 m have been investigated. The reflectarray, due to its flat surface with mechanical advantages and the lack of power divider loss, is one of the strong contenders for large aperture applications. However, when the electrical size increases, as explained in Chapter 4, the bandwidth of the reflectarray becomes very limited. Even with the broadband technique of using the multiple stacked-patch approach as discussed in Chapter 5, the bandwidth cannot be increased enough to satisfy these very large aperture systems. One feasible solution is to use a locally flat but globally parabolic approach also as described in Chapter 5.

A recent development that is worth mentioning is a reflectarray having a rectangular aperture intended for the NASA/JPL's Wide Swath Ocean Altimeter (WSOA) Ku-band radar application. This dual-polarized reflectarray, with an aperture size of $2\,m \times 0.5\,m$, uses variable-size patches as elements. Although this aperture size is not considered very large, its configuration should be employed for very large aperture applications. The required rectangular aperture, as shown in Fig. 7.58, consists of five flat reflectarray subapertures that are connected together to form a curved parabola [51].

The curving of the long dimension of the rectangular surface serves two goals. One is to minimize the incident angles from the feed to the edge elements and, thus, to avoid suffering the losses of the radiation efficiency for these elements. The radiation efficiency discussed here indicates the measure of the amount of energy of each element that is reradiated in the desired main beam direction. The other goal is to maintain a robust bandwidth with minimum time-delay problem that generally becomes more pronounced with a large flat aperture. One key advantage of using reflectarray with flat subapertures is to allow mechanically folding of the flat panels into a compact structure for spacecraft launch-vehicle stowage. Test data indicate that this reflectarray is functioning properly, with good radiation patterns and an aperture efficiency of about 50 percent. Both measured and calculated patterns are given in Fig. 7.59 for the H-polarized field. For the V-polarized field, the patterns are of similar performance and not repeated here.

The antenna configuration just shown is a rectangular aperture with parabolic surface formed only in one dimension. For a very large circular or square aperture, the parabolic shape needs to be formed in two dimensions. This can be done, as shown in Fig. 7.60, by using hexagonal subapertures with each subaperture being a flat reflectarray. As an example, for a 10-m aperture

Figure 7.58. Photos of the piece-wise flat globally parabolic reflectarray (2 m × 0.5 m) for space application.

antenna, each subaperture would be a 3.3-m flat hexagonal reflectarray. This is relatively simple to realize mechanically, since all seven hexagonal subapertures are identical mechanically. Precision hinge and locking mechanisms are needed to deploy the subapertures to form a correct approximation to a parabolic surface. This global parabolic aperture would certainly solve the bandwidth and time-delay issues associated with very large aperture reflectarrays.

7.8 BEAM SCANNING REFLECTARRAYS

One of the key advantages of the reflectarray antenna is its ability to achieve beam scanning by implanting a low-loss phase shifter into each of its elements,

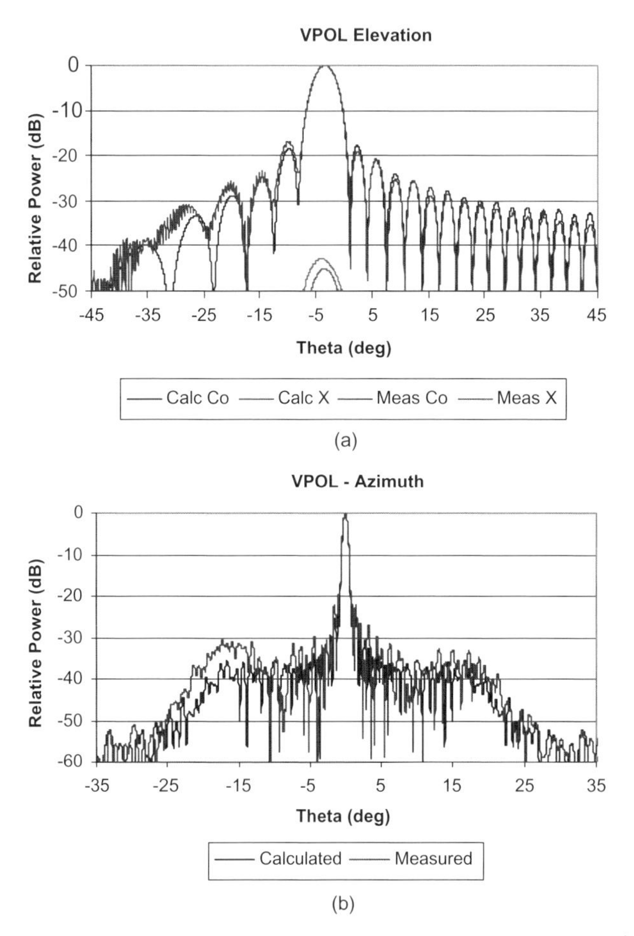

Figure 7.59. Measured and calculated H-polarization patterns in the (a) elevation plane and (b) azimuth plane.

as shown in Fig. 7.61, without the need for expensive transmit/receive (T/R) modules. Conventional phased arrays with large apertures (e.g., elements more than a thousand) will always need T/R modules due to its very lossy and complex power division circuitry. On the other hand, because of the spatial feed of the reflectarray, there is no need for a physical power division circuit and, hence, no T/R module is needed for the reflectarray as long as the phase shifter has relatively low loss. Since a reflectarray generally involves a large number of elements (more than a thousand) and due to a phase front averaging effect, switched-line phase shifters with low bit quantization, such as 2 or 3 bits, are adequate enough to achieve good far-field beam scanning resolution with low sidelobe level if desired.

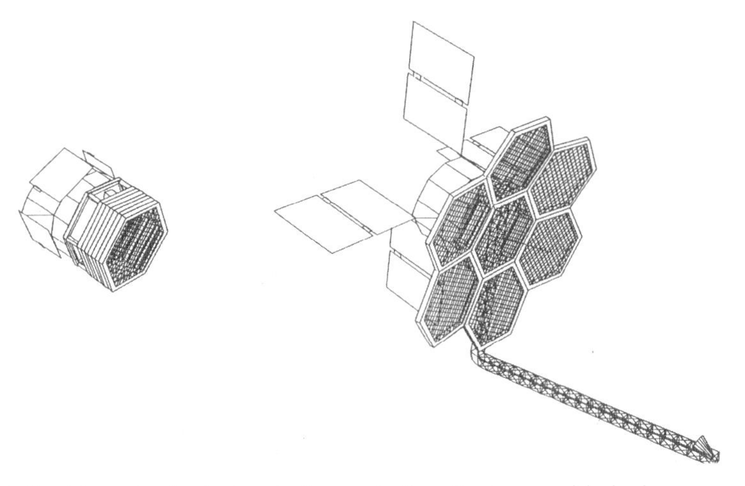

Figure 7.60. Piecewise flat but globally parabolic reflectarray with deployment mechanism for very large aperture application.

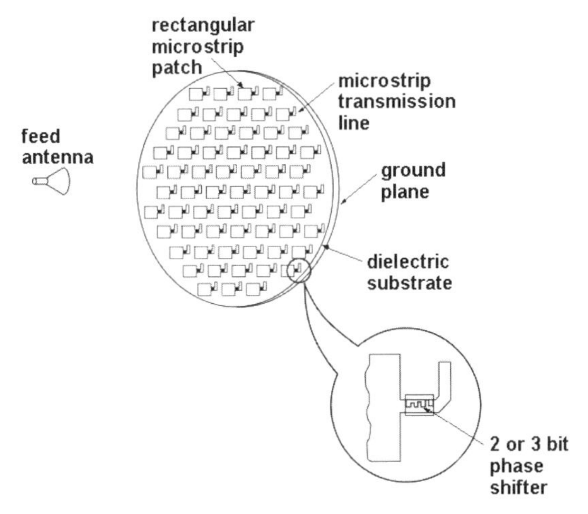

Figure 7.61. Configuration of an electronic beam scanning reflectarray having each element implanted with a phase shifter.

As illustrated in Fig. 7.62 for the reflectarray application, the two- or three-bit switched-line phase shifters are designed with the reflection mode rather than the conventional transmit-through mode. Conventional PIN diodes may be used for these low-bit phase shifters for frequencies below X-band. For frequencies in X-band or above, the low-loss micro-electro-mechanical (MEM)

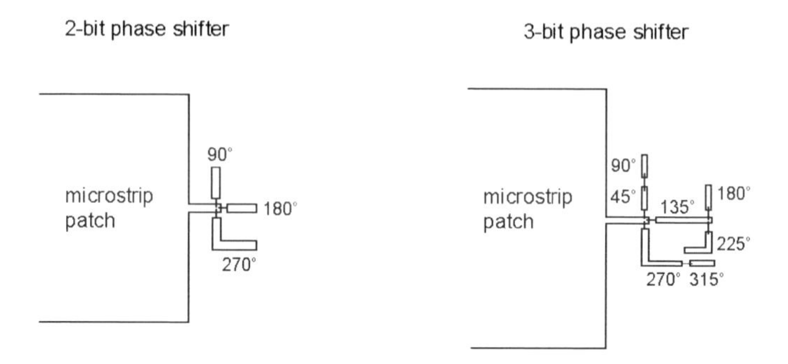

Figure 7.62. Quantized switch-line phase shifters used in reflection mode for reflectarray application.

switch may be used. Low-loss waveguide/ferrite type phase shifter can certainly be used also. The amount of maximum phase shifter insertion loss that can be tolerated for a reflectarray, depending on the application, is generally in the range of one to two decibels.

In the 1996 Phased Array Conference, two beam scanning reflectarrays were reported. One is a 94-GHz monolithic reflectarray [52] fabricated in a single silicon waffle with one-bit PIN diode phase shifters, which achieved ±45° beam scanning. The other is a 35-GHz reflectarray [53] using three-bit waveguide/ferrite phase shifters to achieve ±25° of beam scanning. Both reflectarrays were briefly described in Chapter 2 and are not repeated here. Another beam scanning reflectarray is currently being developed by the NASA Glenn Research Center. It uses a ferroelectric thin film device, $Ba_{0.60}Sr_{0.40}TiO_3$ (BST), for the phase shifter [54] in order to achieve small size. A 19-GHz reflectarray with 615 elements using ferroelectric phase shifters is shown in Fig. 7.63. The ferroelectric phase shifter achieved about 2 dB of loss at the C-band frequency; its loss at 19 GHz is more than 4 dB. A current research effort is being carried out to reduce the insertion loss in the millimeter-wave spectrum.

A more recent development that is worth mentioning here is the use of voltage-controlled Varactor diodes [55], where only one or two control lines are needed to achieve beam scanning with relatively lower loss. The reflectarray configuration is shown in Fig. 7.64, where each of its element is a rectangular patch with two halves connected at the two edges by two varactor diodes. Each diode is voltage controlled to vary its capacitance and hence change the patch's electrical phase. Thus, only one or two control lines are needed per phase shifter. A conventional three-bit switch-line phase shifter would require 12 diodes and 12 control lines. For a reflectarray with thousands of elements, the number of phase shifter control lines becomes a burden for the hardware and should be kept at a minimum.

Figure 7.63. Electronic beam scanning reflectarray at 19 GHz using ferroelectric thin film phase shifters. (Courtesy of Dr. Robert Romanofski of NASA Glenn Research Center.)

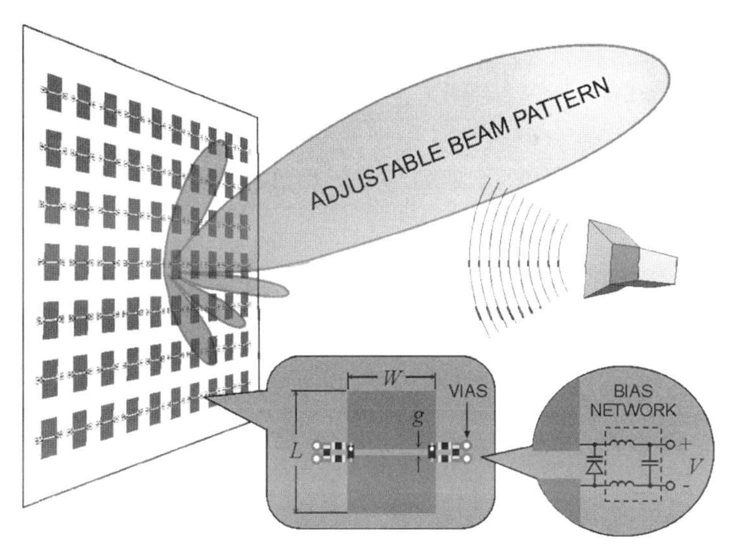

Figure 7.64. Configuration of the beam scanning reflectarray having patch element with two edges loaded with tunable varactor diodes.

For laboratory demonstration, a 70-element C-band reflectarray using varactor diodes was fabricated as shown in the photo of Fig. 7.65. A set of measured scanning beam patterns with $\pm 50^\circ$ scanning range is shown in Fig. 7.66. The loss of the Varactor diode at the C-band frequency is about 2 dB and

Figure 7.65. Photo of the 70-element C-band beam scanning reflectarray with tunable varactor diodes. (Coutesy of Dr. Sean Hum and Prof. Michal Okoniewski of Univ. of Calgary, Canada.)

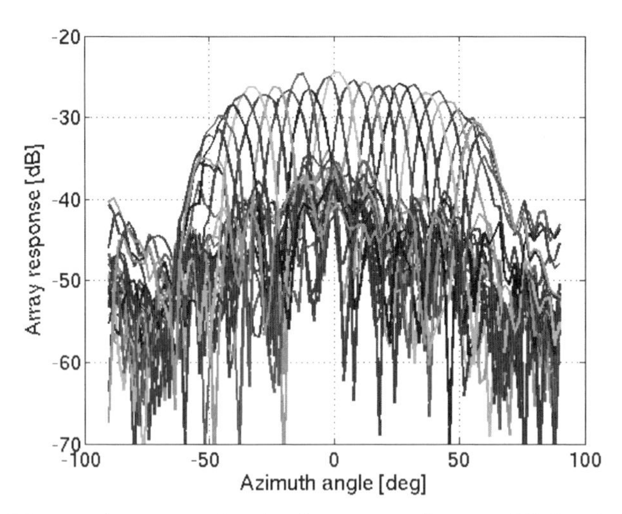

Figure 7.66. Measured patterns showing beam scanning to ±50° for the reflectarray using tunable varactor diodes.

this loss may be significantly increased at the millimeter-wave frequencies. In addition, the diode's packaging issue may limit its use at the higher frequencies. As a result, tuning capacitors based on low-loss micro-electro-mechanical system (MEMS) is currently being investigated for Ka-band application [56].

One proposed low-loss scan technique [44], which has been briefly presented in Chapter 2, is to utilize the angular rotation technique with circularly polarized elements. As depicted in Fig. 2.7, miniature or micro-machined motors could be placed under each element of a reflectarray to achieve wide-angle beam scanning without the need of expensive T/R modules and phase shifters. With this approach, nearly all the major RF losses are eliminated. This concept was later experimentally demonstrated by placing a two-element array with miniature motors in a waveguide simulator [57] and by a small five-element array [58]. To make this concept successful, a controllable stepper miniature motor must be developed so that the motor can be stopped accurately at a set of quantized angular positions. Although the speed of this beam scanning technique is in the millisecond range, rather than the microseconds achievable by electronic beam scanning arrays, for spacecraft applications, the relative motion between the two communication ends often do not require fast beam scanning.

REFERENCES

1. R. A. Russell, T. G. Campbell, and R. E. Freeland, "A technology development program for large space antennas," 31st Congress of the International Astronautical Federation, Tokyo, Japan, Sept. 1980.
2. J. Huang and A. Feria, "A one-meter X-band inflatable reflectarray antenna," *Microwave & Optical Technology Letters*, Vol. 20, Jan. 1999.
3. J. Huang and A. Feria, "Inflatable microstrip reflectarray antennas at X and Ka-band frequencies," IEEE AP-S Symposium, Orlando, Florida, pp. 1670–1673, July 1999.
4. J. Huang and R. J. Pogorzelski, "A Ka-band microstrip reflectarray with elements having variable rotation angles," *IEEE Trans. Antennas and Propagat*, Vol. 46, pp. 650-656, May 1998.
5. J. Huang, V. A. Feria, and H. Fang, "Improvement of the three-meter Ka-band inflatable reflectarray antenna," IEEE AP-S/URSI symposium, Boston, Massachusetts, pp. 122–125, July 2001.
6. H. Fang, M. Lou, J. Huang, L. Hsia, and G. Kerdanyan, "An inflatable/self-regidizable structure for the reflectarray antenna," 10th European Electromagnetics Structure conference, Munich, Germany, Oct. 2001.
7. D.-C. Chang and M.-C. Huang. "Feasibility study of erecting cosecant pattern by planar microstrip reflectarray antenna," AMPC'93, Vol. 2 pp. 19.20–24, 1993.
8. D. M. Pozar, S. D. Targonski, and R. Pokuls, "A shaped-beam microstrip patch reflectarray," *IEEE Transactions on Antennas and Propagation*, pp. 1167–1173, July 1999.

9. J. A. Encinar, et al., "Breadboard of a three-layer printed reflectarray for dual polarisation and dual coverage," 28th ESA Antenna Workshop on Space Antenna Systems and Technologies, Noordwijk (The Netherlands), 30 May–3 June, 2005.
10. J. A. Encinar and J. A. Zornoza, "Three-Layer Printed Reflectarrays for Contoured Beam Space Applications," *IEEE Trans. Antennas Propagat.*, Vol. 52, No. 5, pp. 1138–1148, May 2004.
11. A. Trastoy, F. Ares, and E. Moreno, "Phase-only control of antenna sum and shaped patterns through null perturbation," *Antennas and Propagation Magazine*, pp. 45–54, Dec. 2001.
12. G. M. Kautz, "Phase-only shaped beam synthesis via technique of approximated beam addition," *IEEE Trans. Antennas Propagat.*, Vol. 47, No. 5, pp. 887–894, May 1999.
13. O. M. Bucci and G. D'Elia, "Power synthesis of reconfigurable conformal arrays with phase-only control," *IEE Proc.-Microw. Antennas Propagat.*, Vol. 145, No. 1, pp. 131–136, Feb. 1998.
14. O. M. Bucci, G. Franceschetti, G. Mazzarella, and G. Panariello, "Intersection approach to array pattern synthesis," IEE Proceedings, Vol. 137, pt. H, No. 6, pp. 349–357, December 1990.
15. J. A. Zornoza and J. A. Encinar, "Efficient phase-only synthesis of contoured-beam patterns for very large reflectarrays," *International Journal of RF and Microwave Computer-Aided Engineering*, Vol. 14, pp. 415–423, September 2004.
16. S. Costanzo, F. Venneri, G. Di Massa, and G. Angiulli, "Synthesis of microstrip reflectarrays as planar scatterers for SAR interferometry," *Electronic Letters*, Vol. 39, Issue 3, 6, pp. 266–267, Feb. 2003.
17. J. A. Zornoza and J. A. Encinar, "Multi-layer printed reflectarrays as an alternative to shaped reflectors," in 24th ESTEC Antenna Workshop on Innovative Periodic Antennas, Noordwijk, Holland, pp. 243–247, May 2001.
18. L. I. Vaskelainen, "Phase synthesis of conformal array antennas," *IEEE Trans. Antennas Propagat.*, Vol. 48, No. 6, pp. 987–991, June 2000.
19. J. A. Zornoza and J. A. Encinar, "Design of shaped beam reflectarrays for direct broadcast satellites," in Intl. Symp. on Antennas JINA 2002, Nice, France, pp. 367–370, Nov. 2002.
20. J. A. Encinar, L. Datashvili, J. Agustín Zornoza, M. Arrebola, M. Sierra-Castañer, J. L. Besada, H. Baier, and H. Legay, "Dual-polarization dual-coverage reflectarray for space applications," *IEEE Trans. on Antennas and Propagat.*, Vol. 54, pp. 2827–2837, Oct. 2006.
21. J. A. Zornoza and J. A. Encinar, "Reflectarray pattern synthesis with phase constraints," 13e Journées Internationales de Nice sur les Antennes (JINA), Nice, France, pp. 200–201, Nov. 2004.
22. J. A. Encinar and J. A. Zornoza, "Broadband design of three-layer printed reflectarrays," *IEEE Trans. Antennas Propagat.*, Vol. 51, no. 7, pp. 1662–1664, July 2003.
23. J. A. Zornoza, M. Arrebola, and J. A. Encinar, "Multi-frequency pattern synthesis for contoured beam reflectarrays," in Proc. of 26th Antenna workshop on Satellite Antenna Modeling and Design Tools, ESTEC, Noordwijk, The Netherlands, pp. 337–342, Nov. 2003.

24. M. J. A. Encinar, M. Arrebola, M. Dejus and C. Jouve, "Design of a 1-metre reflectarray for DBS application with 15 percent bandwidth," European Conference on Antennas and Propagation (EuCAP 2006), Nice, France, Nov. 2006.
25. C. J. Sletten, *Reflector and Lens Antennas*, Chapter 7, Artech House 1988.
26. M. Arrebola, J. A. Encinar, Y. Alvarez, and F. Las-Heras, "Design and evaluation of a three-beam LMDS central station antenna using reflectarrays," 13th IEEE Mediterranean Electrotechnical Conference (MELECOM'2006), Benalmádena (Málaga), Spain, May 2006.
27. M. Arrebola, J. A. Encinar, Y. Álvarez, F. Las-Heras, and W. Menzel, "Design of a reflectarray with three shaped beams using the near-field radiated by the feeds," European Conference on Antennas and Propagation (EuCAP 2006), Nice, France, Nov. 2006.
28. J. A. Encinar, "Design of two-layer printed reflectarrays using patches of variable size," *IEEE Trans. Antennas Propagat.*, Vol. 49, No. 10, pp. 1403–14010, Oct. 2001.
29. M. E. Bialkowski, A. W. Robinson, and H. J. Song, "Design, development, and testing of X-band amplifying reflectarrays," *IEEE Trans. Antennas Propag.*, Vol. 50, pp. 1065–1076, August 2002.
30. D. Pilz and W. Menzel, "Folded reflectarray antenna," *Electronics Letters*, pp. 832–833, April 1998.
31. W. Menzel, D. Pilz, and R. Leberer, "A 77-GHz FM/CW radar front-end with a low-profile low-loss printed antenna," *IEEE Trans. Microwave Theory Tech.*, Vol. 47, No. 12, pp. 2237–2241, Dec. 1999.
32. W. Menzel, D. Pilz, and M. Al-Tikriti, "Millimeter-wave folded reflector antenna with high gain, low loss, and low profile," *IEEE AP Magazine*, pp. 24–29, June 2002.
33. W. Menzel, D. Pilz, and M. Al-Tikriti, "A 60 GHz triple folded reflector antenna," *Electronics Letters*, pp. 1075–1076, Sept. 2002.
34. W. Menzel, M. Al-Tikriti, and R. Leberer, "A 76 GHz multiple-beam planar reflector antenna," European Microw. Conf., Milano, Italy, Sept. 2002, pp. 977–980, (vol. III).
35. P. S. Holt and A. Mayer, "A design procedure for dielectric microwave lenses of large aperture ratio and large scanning angle," IRE Trans. on AP-5 (1957), pp. 25–30.
36. A. Y. Niazi and P. J. Mitchell, "Millimetre wave phase corrected reflector antenna," IEE International Conference on Antennas and Propagation, ICAP'83, Norwich, April 1983, Pt. 1, pp. 51–54.
37. W. Menzel and R. Leberer, "Folded reflectarray antennas for shaped beam applications", European Conference on Antennas and Propagation (EuCAP 2006), Nov. 2006.
38. A. Chakraborty, B. N. Das, and G. S. Sanyal, "Beam shaping using nonlinear phase distribution in a uniformly spaced array," *IEEE Trans. Antennas Propagat.*, Vol. AP-30, no. 5, pp. 1031–1034, Sept. 1982.
39. M. Thiel and W. Menzel, "A multiple-beam sector antenna with a dual planar reflectarray arrangement," Europ. Microw. Conf. 2006, Sept. 2006, Manchester, UK.
40. R. Leberer and W. Menzel, "A dual planar reflectarray with synthesized phase and amplitude distribution," Propagat., pp. 3534–3539, Nov. 2005.

41. J. A. Zornoza, R. Lebber, J. A. Encinar, and W. Menzel, "Folded multi-layer microstrip reflectarray with shaped pattern," *IEEE Trans. on Antennas and Propag.*, Vol. 54, No. 2, pp. 510–518, Feb. 2006.
42. P. W. Hannan, "Microwave antennas derived from the Cassegrain telescope," *IRE Trans. Antennas Propagat.*, Vol. AP-9, March 1961, pp. 140–153.
43. B. Houshmand, S. W. Lee, Y. Rahmat-Samii, and P. T. Lam, "Analysis of near-field Cassegrain reflector: Plane wave versus element-by-element approach," *IEEE Trans. Antennas Propagat.*, Vol. 38, July 1990, pp. 1010–1017.
44. J. Huang, "Bandwidth study of microstrip reflectarray and a novel phased reflectarray concept," IEEE AP-S/URSI Symposium, Newport Beach, California, June 1995, pp. 582–585.
45. J. Huang and J. R. Lesh, "Reflectarray elements on sub-reflector for fine beam pointing," JPL New Technology Report no. 41562, Dec. 2004.
46. D. M. Pozar, S. D. Targonski, and H. D. Syrigos, "Design of millimeter-wave microstrip reflectarrays," *IEEE Trans. Antennas Propagat.*, Vol. 45, Feb. 1997, pp. 287–296.
47. J. Huang, C. Han, and K. Chang, "A Cassegrain offset-fed dual-band reflectarray," IEEE AP-S/URSI Symposium, Albuquerque, New Mexico, July 2006, pp. 2439–2442.
48. B. Khayatian and Y. Rahmat-Samii, "Characterizing reflectarray antennas radiation performance," IEEE AP-S/URSI Symposium, Columbus, Ohio, June 2003, pp. III 298–301.
49. K. Mahadevan, S. Ghosh, R. Mizzoni, and G. Martirano, "Precision analysis and design of a triple band feed for the high gain antenna of Cassini deep space mission to Saturn," JINA conference, Nice, France, Nov. 2005.
50. R. L. Jordan, B. L. Huneycutt, and M. Werner, "The SIR-C/X-SAR synthetic aperture radar system," *IEEE Trans Geoscience and Remote Sensing*, Vol. 33, July 1995, pp. 829–839.
51. R. Hodges and M. Zawadzki, "Design of a large dual polarized Ku-band reflectarray for spaceborne radar altimeter," IEEE AP-S Symposium, Monterey, California, June 2005, pp. 4356–4359.
52. J. M. Colin, "Phased array radars in France: present and future," IEEE symposium on Phased Array System and Technology, Boston, Massachusetts, October 1996, pp. 458–462.
53. A. A. Tolkachev, V. V. Denisenko, A. V. Shishlov, and A. G. Shubov, "High-gain antenna system for millimeter-wave radars with combined electrical and mechanical beam steering," IEEE symposium on Phased Array System and Technology, Boston, Massachusetts, October 1996, pp. 266–271.
54. F. Xiong and R. Romanofsky, "Study of the behavior of digital modulations of beam steerable reflectarray antennas," *IEEE Trans. Antennas Propagat.*, March 2005, pp. 1083–1097.
55. S. V. Hum and M. Okoniewski, "An electronically tunable reflectarray using varactor-diode-tuned elements," IEEE AP-S/URSI Symposium, Monterey, California, June 2004, vol. 2, pp. 1827–1830.
56. S. V. Hum, G. McFeetors, and M. Okoniewski, "A reflectarray cell based on a tunable MEMS capacitor," IEEE AP-S/URSI Symposium, Albuquerque, New Mexico, July 2006, URSI session 458.

57. A. Martynyuk, J. Lopez, J. Cuevas, and Y. Sydoruk, "Wideband reflective array based on loaded metal rings," IEEE MTT-S Microwave symposium, Long Beach, California, June 2005.
58. V. F. Fusco, "Mechanical beam scanning reflectarray," *IEEE Trans. Antennas Propagat.*, Vol. 53, Nov. 2005, pp. 3842–3844.

INDEX

Alternating projection algorithm, for contoured-beam reflectarrays, 148–149
Amplifying reflectarrays, 177–182
Analysis tool, for Cassegrain offset-fed reflectarray, 193
Analysis techniques, 27–74. *See also* Method of Moments (MoM) technique
  Characteristic Basis Functions, 32
  CPU time in, 32–33
  design curves, 29–30
  Fast Fourier Transforms, 32
  feed model and radiation patterns in, 64–74
  Finite Difference Time Domain modeling, 31
  Finite Element Method, 31
  full-wave, 30
  Generalized Scattering Matrix, 39–40, 42–44, 44–46
  for multilayer periodic structures, 38–48
  overview of, 29–34
  phase-shift distribution, 33–34, 34–35
  for phase-shifter elements based on aperture-coupled patches, 33, 39, 55–64, 65
  for phase-shifter elements based on variable-sized patches, 31–32, 33, 39, 48–55
  for rectangular patches with attached stubs, 31, 35–38
  reflectarray elements in, 30–31
  Sparse-Matrix Canonical Grid Method, 32
Angular rotation technique, 15–16
Annular rings
  for dual-band multilayer reflectarrays, 18
  low-frequency above high-frequency, 128–131
  printed, 13, 14
Antennas. *See also* Parabolic antennas; Reflectarray antenna
  Astro mesh, 137
  bifocal, 184–185
  high-gain array, 1
  inflatable, 16
Aperture-coupled patches, 28–29
  broadband, 96–97
  design of, 56–57
  impedance matching in, 56–57, 94
  phase delay curves for, 56, 57–64, 65
  phase shift and losses in, 57–64, 65
  phase-shifter elements based on, 33, 39, 55–64, 65
  true time delay with, 106, 107, 108
Aperture efficiency, 88–91
Aperture electrical size, 88–91
Approximation methods, in radiation pattern calculation, 86
Array equation for grating-lobe avoidance, 84
Array lens, 20–22
Array summation technique, in radiation pattern calculation, 85–86
Astro mesh antenna, 137
Attached stubs, 28–29
  analysis of rectangular patches with, 31, 35–38, 39

Reflectarray Antennas, by John Huang and José A. Encinar

Bandwidth. *See also* Broadband techniques; Dual-band reflectarrays
of reflectarray antennas, ix, x–xi, 4–6
of reflectarray elements, 82
Bandwidth improvement, 99–100
for aperture-coupled patches, 96–97
for stacked variable-sized patches, 97–99
Bandwidth limitation
by differential spatial phase delay, 4–6, 100–104, 105
by reflectarray elements, 93–95
Basis functions, subdomain, 32
Beam scanning reflectarrays, 195–201
Beamwidth
of multi-beam reflectarrays, 174–175
of reflectarray elements, 81–82
Bifocal antennas, 184–185
Breadboards
for DBS transmit antenna, 155–156, 159, 162–165
for two-beam reflectarray, 172
Broadband phase-shifter elements, 95–100
Broadband techniques, 93–116
for large reflectarrays, 104–116
Building block, periodic interface as, 40–44

Carbon–fiber–reinforced plastic (CFRP) layer, 45
in contoured-beam reflectarrays, 157
Carpenter tapes, for X-band inflatable reflectarray, 142
Cascade process, 45
Cassegrain center-fed reflectarrays, 190–191
Cassegrain configuration, for amplifying reflectarray, 181–182
Cassegrain offset-fed X/Ka dual-band reflectarray, 128–131
Cassegrain offset-fed reflectarrays, 190–193
Cassegrain offset-feed system, 18
C-band reflectarray, 198, 199, 200
Center-fed Cassegrain configuration, 190–191
Characteristic Basis Functions (CBFs), 32
Circularly polarized microstrip patches, 13
Circularly polarized phased array, 13
Circular polarization (CP), 28, 31
with annular ring elements, 128–131
angular rotation technique, 15–16, 201
element reflection efficiency and, 83
with two closely spaced frequencies, 120–121, 122
with two widely spaced frequencies, 119–120, 121
Circular reflectarray, phase-shift distribution in, 34–35
Circular two-layer reflectarray, 72–73
Compensating phase-shift, 101
Compensating spatial phase delay. *See also* Differential spatial phase delay
in a frequency band, 108–112
for large reflectarrays, 105
Concavely curved reflectarrays, 5
Contoured-beam reflectarrays, 143–169
practical designs for, 150–169
Coordinate systems, for contoured-beam reflectarrays, 145–147
Cophasal far-field reflectarray beam, 13
Coverage requirements, for contoured-beam reflectarrays, 155–169
CPU times, in reflectarray analysis, 32–33
Crossed dipoles, 49
with dual-band reflectarrays, 120, 121
low-frequency, 125–127
Cross-polarization
in aperture-coupled patches, 58
in contoured-beam reflectarray design, 151–152, 159–161, 162–165, 168–169
in folded compact reflectarrays, 184
in multiband reflectarrays, 132–134
predicting, 30
in radiation patterns, 71
Cross-polarization performance, of microstrip antennas, 12

Deep-space telecommunications, 125–127, 128–131
  inflatable/thin-membrane reflectarrays for, 137–143
  very large aperture reflectarrays for, 194–195, 196, 197
Delay lines
  with dual-band reflectarrays, 119–120, 121
  for large reflectarrays, 105
  true time delay with, 106–108
Design curves, in reflectarray analysis and design, 29–30
Differential spatial phase delay, bandwidth limitation by, 4–6, 100–104, 105. *See also* Compensating spatial phase delay
Diodes. *See* PIN diode phase shifters; Varactor diodes
Dipole elements, 121–123. *See also* Crossed dipoles; Microstrip dipoles
  in multiband reflectarrays, 131–134
  printed variable-length, 13, 14, 16
Dipole lengths, for microstrip antennas, 12
Direct Broadcast Satellites (DBSs), ix, 143–144
  contoured-beam reflectarray designs for, 150–169
Directivity, computing, 73–74
Direct synthesis method, for contoured-beam reflectarray design, 144
Dissipative losses
  in aperture-coupled patches, 57–64, 65
  in waveguide simulator, 52–55
Distributed grating lobes, 84
Double-layer structures, for bandwidth improvement, 100
Dual-band multilayer reflectarrays, 18
Dual-band reflectarrays, 119–131
  with single-layer substrate, 119–123
  with two-layer substrates, 123–131
Dual coverage, reflectarray geometry for, 155
Dual-gridded reflector, 155
Dual-linear polarization, 28
  with two widely spaced frequencies, 119–120, 121
Dual-linear polarization reflectarrays, 55
Dual-reflector configuration, 190

Earth remote sensing, inflatable/thin-membrane reflectarrays for, 137–143
Efficiency. *See* Aperture efficiency; Element reflection efficiency; Illumination efficiency; Spillover efficiency
Electromagnetic coupling, for bandwidth improvement, 100
Electromagnetic simulators, 33
Element beamwidth, 81–82
Element reflection efficiency, 82–83
Element reflection phase, 79–81
Element selection, 79–84
Element spacing, 83–84
Ellipsoid subreflector, in Gregorian reflectarray, 190
Elliptical reflectarrays, 150–151
Entire-domain basis functions, 43
European coverage. *See* Horizontal (H-) polarization

Faceted configuration, for large reflectarrays, 105–106
Faraday phase shifters, 15
Far-field radiation patterns, calculation of, 64–72, 85–86
Fast beam scanning, by beam scanning reflectarray, 201
Fast Fourier Transforms (FFTs), 32
Feed coordinate system, 67–68
Feed horn. *See also* Gaussian horn antenna
  for Cassegrain offset-fed reflectarray, 190, 191
  for X-band inflatable reflectarray, 140, 141
Feed model. *See also* Offset feed system
  in analysis techniques, 64–74
  for folded compact reflectarrays, 182, 184
  for multi-beam reflectarrays, 170–172, 173–177, 178, 179
Ferrite phase shifters, three-bit, 15
Ferroelectric phase shifter, for beam scanning reflectarray, 198, 199

Finite Difference Time Domain (FDTD) modeling technique, 31
Finite Element Method (FEM), 31
Fletcher Powell algorithm, 110–111
Floquet space harmonics, 39–40, 41–44, 44–48
 in aperture-coupled patch design, 57
 in radiation patterns, 71–72
Floquet's theorem, 31
Folded compact reflectarrays, 182–189
 three-layer, 187, 188
Folded reflectarray configuration, 18–19
Fourier transforms, 68. *See also* Fast Fourier Transforms (FFTs); 2D Inverse Discrete Fourier Transform (IDFT2)
Frequency band, compensation of phase delay in, 108–112. *See also* Multiband reflectarrays; Two-frequency bands
Frequency excursion error, 5
Frequency scanned grating-reflector antenna, 13, 14
Frequency Selective Surfaces (FSSs), 33–34, 125
Fresnel-Zone configuration, 13, 14
Fresnel-Zone plate reflector (FZPR), 20, 22–23
Full-wave analysis/technique
 of multilayer periodic structures, 38–48
 in reflectarray analysis and design, 30–34

Gain, computing, 73–74
Gain requirements
 for contoured-beam reflectarrays, 145–147, 152–154, 159–162, 165, 166
 for multi-beam reflectarrays, 174
Galerkin testing procedure, 43
Galileo spacecraft, 138
Gaussian horn antenna, 165. *See also* Feed horn
Generalized Scattering Matrix (GSM), 31, 39–40, 42–44, 44–46
 in aperture-coupled patch design, 57
Genetic Algorithm (GA) optimization, 16–17
Geometrical Theory of Diffraction, 86. *See also* Uniform Geometrical Theory of Diffraction (UGTD)
Geometry design
 for bandwidth improvement, 99–100
 for reflectarray antenna, 86–91
Grating lobes, 84
Grating-reflector antenna, frequency scanned, 13, 14
Green's functions, 43–44
Gregorian configuration, 190
Ground plane, ohmic losses, 44

Hexagonal subapertures, in very large aperture reflectarrays, 194–195, 197
HFSS simulator, 33
High-frequency annular rings, 128–131
High-frequency elements
 above low-frequency elements, 124–125
 below low-frequency elements, 125–131
High-gain array antennas, 1
Hoop-column umbrella antenna, 137, 138
Horizontal (H-) polarization
 in contoured-beam reflectarrays, 155–169
 in very large aperture reflectarrays, 194, 196
Horn antenna. *See also* Feed horn
 for amplifying reflectarrays, 178
 Gaussian, 165
 for multi-beam reflectarrays, 177
H-wall waveguide simulator, for phase calibration, 80
Hyperboloid subreflector, in Cassegrain reflectarray, 190, 191, 192–193

Illumination efficiency, 87–88
Impedance, modal, 46
Impedance matching, 56–57, 93–94
Incident field, 41, 42, 66–68
Indium-tin-oxide (ITO) film, 19–20
Infinite-array approach, 31–32
 for phase calibration, 79–80
 in spatial phase delay compensation, 109–110
Inflatable antenna, 16, 129, 137–143

Infrared patch reflectarray, 20
Intersection Approach technique
  for contoured-beam reflectarrays, 145, 147–150, 157, 166, 169
  for folded compact reflectarrays, 187
  for multi-beam reflectarrays, 174–175
Inverse Discrete Fourier Transform (IDFT2), 70

Ka-band 3-m inflatable reflectarray, 139–143
Ka-band elements, 125–127, 128–131
Ka-band membrane, in Cassegrain offset-fed reflectarray, 192–193
Ku-band reflectarrays, 143–144
  very large aperture, 194

Large reflectarrays. *See also* Very large aperture reflectarrays
  bandwidth limitation in, 100–104, 105
  broadband techniques for, 104–116
Left-hand circular polarization (LHCP), 15
Linear polarization
  with two closely spaced frequencies, 123
  with two widely spaced frequencies, 121–123
LMDS (Local Multipoint Distribution Service) central station antenna, 178, 187–189
Losses. *See also* Low-loss scan technique
  in aperture-coupled patches, 57–64, 65
  in waveguide simulator, 52–55
Low-frequency annular rings, 128–131
Low-frequency elements
  above high-frequency elements, 125–131
  below high-frequency elements, 124–125
Low-loss scan technique, for beam scanning reflectarray, 201
Low-profile printed antennas, development of, ix
Ludwig definition, 71

Masks
  for contoured-beam reflectarrays, 146–147, 147–148, 155
  photo-etching, 155
Membranes. *See* Thin membranes
MEMS (micro-electro-mechanical system), for Ka-band applications, 201. *See also* Micro-electro-mechanical (MEM) switches
MESFET technology, with amplifying reflectarrays, 179
Method of Moments (MoM) technique, 12–13, 31–33, 36, 38–43. *See also* Spectral Domain Method of Moments (SD-MoM)
  for multi-beam reflectarrays, 177, 178, 179
  in radiation pattern calculation, 86
  in spatial phase delay compensation, 109–110, 115–116
Micro-electro-mechanical (MEM) switches, 15. *See also* MEMS (micro-electro-mechanical system)
  for beam scanning reflectarray, 197–198
Micro-machined motors, 15, 201
Microstrip amplifying reflectarray, 178–179, 180
Microstrip dipoles, 27, 59, 99, 184
Microstrip patches, 3–4, 27–29, 34–38, 93
  with array lens, 20–21
  circularly polarized, 13
  for Cassegrain offset-fed reflectarray, 191–192
Microstrip reflectarray antenna, 11–13
Miniature motors, 15, 201
Modal impedance, 46
Modular approach
  analysis techniques using, 39–40, 48
  aperture-coupled patch design using, 57
Monolithic reflectarray antenna, 15
Multiband reflectarrays, 131–134
Multi-beam reflectarrays, 169–177, 178, 179
Multifacet reflectarrays, 112–116
Multi-fed reflectarrays, 175–177, 178, 179
  folded, 184

Multilayer elements, 50–52, analysis techniques for, 44–48
Multilayer periodic structures, analysis techniques for, 38–48
Multilayer reflectarrays, dual-band, 18
Multilayer stacked patches, 16, 17, 50–55
Multipanel configuration, 112–113, 114

Offset-feed system. *See also* Feed model
  Cassegrain, 18, 190–193
  for contoured-beam reflectarrays, 165–166
  for X-band inflatable reflectarray, 141
1-D multipanel configuration, 113
One-layer reflectarray. *See* Single entries
Optimization techniques. *See also* Genetic Algorithm (GA) optimization
  in contoured-beam reflectarray design, 152, 157–159, 166
  for large reflectarrays, 111–112
Overall approach, analysis techniques using, 39–40

Parabolic antennas, 1, 3
  very large aperture reflectarrays as, 194–195, 197
Parabolic reflectors, 173–174
  dual-band techniques versus, 119
  multifacet reflectarrays as approximating, 112–116
  space-deployable, 137–138
Parallel-plate waveguide simulator, for phase calibration, 80
Patches. *See also* Infrared patch reflectarray; Microstrip patches; Stacked patches
  aperture-coupled, 28–29, 33, 39, 55–64, 65, 94, 96–97, 106, 107, 108
  high-frequency, 125–127
  multilayer stacked, 16, 17, 50–55
  on periodic interface, 40–44, 72–73
  printed, 27–29, 34–38, 93–94
  rectangular, 28, 31, 35–38, 39, 49–50, 50–52, 100, 106, 107, 108
  ridge-shaped, 99
  single, 48–55
  variable-sized, 31–32, 33, 39, 48–55, 97–99
Patch sizes, for microstrip antennas, 12–13
Path length, calculation of, 84–85
Pattern synthesis, for contoured-beam reflectarrays, 145–150
Periodic interface, as building block, 40–44
Phase calibration
  H-wall waveguide simulator for, 80
  infinite-array approach for, 79–80
  waveguide measurement approach for, 80–81
Phase control, 28
Phase curves, 33, 36–38, 49–52, 57–64, 94–95
Phased array
  circularly polarized, 13
  large aperture, 196, 198
Phased Array Conference of 1996, 15, 198
Phased array feed, for Cassegrain offset-fed reflectarray, 191, 192
Phase delay. *See also* Compensating spatial phase delay
  bandwidth limitation by, 4–6, 100–104, 105
  calculation of, 84–85
  in frequency band, 108–112
Phase delay curves, for aperture-coupled patches, 56, 57–64, 65
Phase delay difference, 110–111
Phase delay stubs, 11–12, 13, 28–29, 57–64, 106–108
Phase difference, 109–110, 111–112
Phase distribution
  for contoured-beam reflectarrays, 148–149, 152, 158
  for folded compact reflectarrays, 186
  for multi-beam reflectarrays, 170–172
Phase-only synthesis, for contoured-beam reflectarrays, 145, 147–149
Phase response, computing, 36–37
Phase-shift, 27–29. *See also* Reflection phase
  in aperture-coupled patches, 57–64, 65
  compensating, 101–102

computing, 47–48, 49–50, 51–52
with folded compact reflectarrays, 182–184, 185, 186
in multi-beam reflectarrays, 169–170
in reflectarray elements, 94–95
in waveguide simulator, 52–55
Phase-shift distribution, 33–34, 34–35
in contoured-beam reflectarray design, 151–152, 157, 158
Phase-shifter elements
based on aperture-coupled patches, 33, 39, 55–64, 65
based on variable-sized patches, 31–32, 33, 39, 48–55
broadband, 95–100
Phase shifters, 1, 3–4. *See also* beam scanning reflectarray, 196, 197–198
Faraday, 15
PIN diodes, 15, 197–198
three-bit ferrite, 15
Phase variation
for contoured-beam reflectarrays, 149–150
smooth, 97
Phase-versus-element-change curve, 81
Photo-etching masks, 155
Photonically controlled reflectarray, 19–20
Physical optics (PO) method, for Cassegrain offset-fed reflectarray, 193
Piecewise flat reflectarrays, 5, 112–116
PIN diode phase shifters, for beam scanning reflectarray, 198
Polarization, 28. *See also* Circularly polarized phased array; Circular polarization (CP); Cross-polarization; Dual-linear polarization; Dual-linear polarization reflectarrays; Left-hand circular polarization (LHCP); Linear polarization; Right-hand circular polarization (RHCP); Single linear polarization reflectarrays
for contoured-beam reflectarrays, 155–169
for folded compact reflectarrays, 182–184, 187
for multi-beam reflectarrays, 174–175, 177, 178, 179
Power handling, for reflectarray antenna, 91
Printable microstrip antennas, 11–13
Printed annular rings, 13, 14
Printed antennas, 3–6, 11–13
development of, ix
Printed circuit technology, 48
Printed patches, impedance matching in, 56–57, 93–94
Printed reflectarray, operating principle of, 27–29. *See also* Reflectarray antenna
Printed-reflectarray elements, with thin membranes, 16
Printed variable-length dipole elements, 13, 14, 16
Proximity-coupled stubs, for bandwidth improvement, 100
Pyramidal horn antenna, for multi-beam reflectarrays, 177

Radar cross-sections (RCSs), 82, 132–134
Radiating elements, in aperture-coupled patch design, 57
Radiation patterns. *See also* Pattern synthesis
from amplifying reflectarrays, 178–179
analysis techniques to determine, 64–74
from beam scanning reflectarrays, 199–201
calculation of, 85–86
from Cassegrain offset-fed reflectarray, 193
from contoured-beam reflectarrays, 144–150, 152–154, 155, 156, 162, 163, 164, 166–169
from folded compact reflectarrays, 185–186, 188, 189
from Ka-band and X-band elements, 125–127, 130–131
from Ka-band inflatable reflectarray, 142–143

from multi-beam reflectarrays, 172–174, 177, 178, 179
from two-layer elements, 97–99
in spatial phase delay compensation, 111–112
from three-panel centered-fed reflectarray, 113, 115, 116
from X-band inflatable reflectarray, 139
Rectangular patches, 28
analysis techniques for, 31, 35–38, 39
with attached stubs, 31, 35–38, 39
for bandwidth improvement, 100
in multilayer structure, 50–52
in single-layer structure, 49–50
with slots, 100
true time delay with, 106, 107, 108
Reflectarray antenna, 1–6. *See also* Large reflectarrays; Printed reflectarray
advantages of, 3–4
amplifying, 177–182
applications of, xi, 137
array lens versus, 20–22
based on two layers of variable-sized patches, 72–73
beam scanning, 195–201
broadband techniques for, 93–116
calculating path length and phase delay in, 84–85
calculating radiation pattern of, 64–72, 85–86
Cassegrain offset-fed, 190–193
Cassegrain offset X/Ka dual-band, 128–131
as Cassegrain reflectarray subreflector, 191, 192
contoured-beam, 143–169
design, configuration, and operation of, x, 1–3
design curves in analysis and design of, 29–30
development of, ix, 9–23
disadvantage of, 4–6
dual-band, 119–131
folded compact, 182–189
Fresnel-Zone plate reflector versus, 20, 22–23
geometry design for, 86–91
with high-frequency elements above low-frequency elements, 124–125
inflatable, 137–143
with low-frequency elements above high-frequency elements, 125–131
microstrip, 11–13
monolithic, 15
multiband, 131–134
multi-beam, 169–177, 178, 179
multifacet, 112–116
multi-fed, 175–177, 178, 179
power handling for, 91
practical design aspects of, 79–91
printed, 3–6
recent and future applications for, 137–201
spatial power combining, 179–181
spiralphase, 9–11
techniques for analyzing, 27–74
thin-membrane, 137–143
true time delay, 106–108
two-beam, 170–173
two-layer, 72–73, 172, 173
very large aperture, 194–195, 196, 197
waveguide, 9
X-band, 16, 17
Reflectarray configuration, folded, 18–19
Reflectarray element pattern effect, 81–82, 182–189
Reflectarray elements, 2, 4. *See also* Element entries; Phase-shifter elements; Printed variable-length dipole elements
analysis and design of, 30–31
bandwidth limitation by, 93–95
design of, 56–57, 79–91
infrared, 20
in periodic environment, 44–48
photonically controlled, 19–20
with solar cells, 18
Reflected waves, from reflectarray elements, 36–38, 49–52, 57–64, 83
Reflection coefficient, 36–38, 49–52, 57–64, 109, 110
Reflection efficiency, of reflectarray elements, 82–83

Reflection phase, of reflectarray elements, 36–38, 49–52, 57–64, 79–81. *See also* Phase entries
Reradiated waves, from reflectarray elements, 83
Resonant dimensions, varying, 28
Ridge-shaped patches, for bandwidth improvement, 99
Right-hand circular polarization (RHCP), 15
Rigid-frame structures, 129
Ring elements. *See* Annular rings
Rohacell® dielectric, 51–52, 54

Sandwich configuration, for contoured-beam reflectarrays, 165–166
Satellites, reflectarray antennas for, 16, 18. *See also* Space entries
Scattered field, 42, 43
Scattering parameters, 39–40
S-curve, of phase-change versus element-change, 49–50, 81
Shaped beam applications, for folded compact reflectarrays, 185–186, 187, 188–189
Shaped beam reflectarrays. *See* Contoured-beam reflectarrays
Single-layer elements
  analysis of, 49–50
  improving bandwidth of, 97–99
Single-layer substrate, dual-band reflectarrays with, 119–123
Single linear polarization reflectarrays, 55
Single patches, phase-shifter element based on, 48–55
Slot-array, with folded compact reflectarrays, 182
Slots, in rectangular patches, 100
Smooth phase variation, 50–52, 97
Solar cells, with reflectarray elements, 18
Spacecraft. *See also* Deep-space telecommunications
  contoured-beam reflectarrays for, 143–169
  reflectarray antennas for, 16, 18
Space-deployable antennas, 137–143
  contoured-beam reflectarray, 143–169
Spacing, of reflectarray elements, 83–84
Sparse-Matrix Canonical Grid Method (SMCG), 32
Spatial power combining reflectarray, 179–181
Spectral Domain Method of Moments (SD-MoM), 31, 32, 33, 36, 42, 43, 45. *See also* Method of Moments (MoM) technique
  in contoured-beam reflectarray design, 151, 157
Spectral functions, 69–71
Spillover efficiency, 87–88
Spiralphase reflectarray antenna, 9–11
Spring-tape reinforced (STR) boom, for X-band inflatable reflectarray, 141–142
Square loops
  for dual-band reflectarrays, 120
  for multiband reflectarrays, 132–134
Stacked array layers, 28
Stacked patches
  dual-band reflectarrays with, 123–124
  multilayer, 16, 17
Stacked variable-sized patches
  broadband, 97–99
  phase-shifter element based on, 48–55
Stubs. *See* Attached stubs; Phase delay stubs; Proximity-coupled stubs
Subdomain basis functions, 32
Summation technique, in radiation pattern calculation, 85–86
Sun-flower antenna, 137
Switches, micro-electro-mechanical, 15
Switching diodes, in spiralphase reflectarray antennas, 9–10
Synthesis technique, for contoured-beam reflectarrays, 147–149
Synthetic Aperture Radar (SAR) antennas, 113, 114
  multi-beam, 174–175

Tangential electric (TE) field. *See* Floquet space harmonics
Tangential magnetic (TM) field. *See* Floquet space harmonics
TE Floquet harmonics. *See* Floquet space harmonics

Thin film devices, for beam scanning reflectarray, 198, 199
Thin membranes, 137–143
  in Cassegrain offset-fed reflectarray, 192–193
  printed-reflectarray elements with, 16
  two-layer, 128–129
Three-bit ferrite phase shifters, 15
Three-layer reflectarrays, 110–112, 150–151
  folded compact, 187, 188–189
Three-panel centered-fed configuration, 113, 114
  radiation patterns from, 113, 115, 116
TM Floquet harmonics. *See* Floquet space harmonics
Transmit/receive (T/R) modules, 196, 201
Tri-band elements, 123
True time delay compensation broadband technique, 101, 106–108
Two-beam reflectarrays, 170–173
Two-dimensional arrays, analysis techniques for, 36
2D Inverse Discrete Fourier Transform (IDFT2), 70
Two-frequency bands, polarization with widely spaced, 119–120, 121
Two-layer reflectarrays, 72–73, 172, 173
  improving bandwidth with, 97–99
Two-layer substrates, dual-band reflectarrays with, 123–131
Two-layer thin membranes, ring elements on, 128–129

Umbrella radial rib mesh reflector antenna, 138
Uniform Geometrical Theory of Diffraction (UGTD), 88–89. *See also* Geometrical Theory of Diffraction
U-shaped delay line, true time delay with, 106–108

Varactor diodes, 15
  for beam scanning reflectarray, 198, 199–201
Variable-sized crossed dipoles, with dual-band reflectarrays, 120, 121
Variable-sized patches, 31–32, 33, 39
  broadband, 97–99
  phase-shifter element based on, 48–55
Vertical (V-) polarization
  for contoured-beam reflectarrays, 155–169
  for very large aperture reflectarrays, 194, 196
Very large aperture reflectarrays, 194–195, 196, 197

Waveguide measurement approach, for phase calibration, 80–81
Waveguide reflectarray antenna, 9
Waveguides, in spiralphase reflectarray antennas, 9–10
Waveguide simulator (WGS), phase shift and losses in, 52–55, 59–64, 65
Wide Swath Ocean Altimeter (WSOA), 113
  very large aperture reflectarrays for, 194
WR90 waveguide, 53
WR112 waveguide, 59, 60
Wrapped-rib antenna, 137, 138

X-band 1-m inflatable reflectarray, 138–139
X-band beam scanning reflectarray, 197
X-band elements, 125–127, 128–131
X-band membrane, in Cassegrain offset-fed reflectarray, 192–193
X-band reflectarray antenna, 16, 17
X-band Synthetic Aperture Radar antennas, 174–175
X-polarization. *See* Cross-polarization